REVEALING THE UNIVERSE

REVEALING THE UNIVERSE

THE MAKING OF THE CHANDRA X-RAY OBSERVATORY

WALLACE TUCKER AND KAREN TUCKER

HARVARD UNIVERSITY PRESS
Cambridge, Massachusetts
London, England
2001

Copyright © 2001 by Wallace Tucker and Karen Tucker

ALL RIGHTS RESERVED

Printed in the United States of America

Library of Congress Cataloging-in-Publication Data

Tucker, Wallace H.
Revealing the universe : the making of the Chandra X-ray Observatory / Wallace Tucker and Karen Tucker.
p. cm.
Includes bibliographical references and index.
ISBN 0-674-00497-3 (hardcover : alk. paper)
1. Chandra X-ray Observatory (U.S.)—History. 2. X-ray astronomy—United States—History.
I. Tucker, Karen. II. Title
QB472.T82 2001
522′.6863—dc21 00-053862

CONTENTS

REVEALING THE UNIVERSE

Introduction

ONE COOL DECEMBER evening in southern California we were attending the board meeting of a local environmental organization. Suddenly, apropos of nothing, one of the group asked, "Can you show us the planets that the media is making such a fuss over?"

We trooped outside onto the lawn, where we were treated to a view of a rare planetary alignment. Mercury, barely visible on the western horizon in the fading light of the setting sun, then Venus, the moon, Mars, Jupiter, and Saturn, were lined up in a graceful arc across the sky. There were the expected oohs and ahs, then, the group's curiosity about the cosmos satisfied, we returned to tackle more down-to-earth matters such as expenses and budget projections. Two hours later, as we left the meeting, Mercury had set, but on that exceptional night we were compensated by the spectacle of thousands of stars and the great white smear of the Milky Way.

THE CELESTIAL OBJECTS we observed that night, plus a few thousand stars that were below the horizon and an occasional comet, constitute what was once believed to be the entire universe. Then about 400 years ago this belief was shattered. A simple instrument had been invented that enabled humankind to see objects too faint or distant to be detected by the human eye. Military minds of the time quickly recognized the potential of the telescope as a tool to spy on the enemy, but its greater significance was appreciated by Galileo. In his hands it became the instrument of an intellectual revolution as it transformed people's view of their place in the grand scheme of things. The moon, thought to be polished and smooth, was seen to be "rough and uneven, just like the face of the Earth itself." Other moons were discovered to be orbiting Jupiter. And the Milky Way was found to be "nothing else but a mass of innumerable stars."

Galileo deciphered two crucial messages from the stars: first, that there is much more to the universe than we can see with the naked eye, and second, that the Earth is not distinct from the universe, but part of it.

In the 390 years since Galileo's discoveries, astronomers have continued to explore the heavens with increasingly sophisticated and powerful optical telescopes located on mountaintops and in orbit around the Earth. We now know that the Milky Way is a galaxy, a vast rotating, gravity-bound system of hundreds of billions of stars, including the sun.

We also know that, just as our sun is not unusual among stars, our Milky Way galaxy is not exceptional among galaxies. The size of the known universe is measured in billions of light years and contains a billion or more galaxies, each of which contain hundreds of billions of stars. The Hubble Space Telescope and other large optical telescopes have revealed a more complex, dazzling, and fantastic universe than we could ever have imagined on the basis of information gathered by the unaided eye.

Alongside these monumental achievements of twentieth-century astronomy there has been yet another revolutionary discovery—that optical telescopes reveal only a portion of the universe. The visible light portion.

Optical telescopes cannot show us the invisible microwaves that fill the dark spaces between galaxies and tell us about the Big Bang from which our universe emerged. Nor can they image the cool giant molecular clouds that reveal, through invisible radio and infrared radiation, clues about the formation of planetary systems such as our solar system. Optical telescopes cannot focus the invisible high-energy X-rays that stream from superheated matter in the warped space near the event horizons of black holes, and that flood from vast intergalactic clouds of multimillion-degree gas that may constitute as much as half the matter in the universe. Nor can they detect the incredibly intense rapid bursts of invisible gamma rays that occur in galaxies 10 billion light years away.

New kinds of telescopes and detectors, built especially for each of these invisible wavelengths of light, reveal essential and unique information about the origin, evolution, and destiny of the universe. Our story will focus on the astronomy of one particular type of light, X-radiation, and on the epic human effort expended to build the Chandra X-ray Observatory. Chandra is a large, extremely sophisticated X-ray telescope that will complement the Hubble Space Telescope as part of NASA's fleet of Great Observatories for the twenty-first century.

During the first 20 or so years this great telescope was being planned and built, it had the name Advanced X-ray Astrophysics Facility, or AXAF. In December 1998, it was renamed Chandra X-ray Observatory in honor of the late Nobel laureate Subrahmanyan Chandrasekhar, widely regarded as one of the foremost astrophysicists of the twentieth century. For reasons of historical accuracy, we will use AXAF when referring to the observatory before the name change.

When the first, rather simple, X-ray detectors revealed many places in the universe where temperatures reach millions of degrees, pioneering X-ray astronomers realized, as did Galileo when he first looked through his telescope, that they were onto something big. They knew that a new kind of telescope, an X-ray telescope, would have to be created. And ultimately, to achieve parity with optical astronomy, they would have to build a large, orbiting X-ray observatory. What they didn't comprehend, in their enthusiasm over having created a new field of exploration, was how long it would take.

The technical challenges encountered in building a telescope able to focus X-rays as precisely as major optical telescopes focus visible light would be immense. The political struggles to secure funding for the project would be grueling. Transforming the gleam in the eye of a visionary X-ray astronomer into a reality would require special talents, special training, and special people. People who could build instruments as well as use instruments. People who could work in teams as well as independently. People who could push rivals aside and bring them together. People who could persist for decades to fulfill their dreams and ambitions.

Yet in another sense, the dream of the Chandra X-ray Observatory has become a reality with amazing speed. The first cosmic X-ray source was discovered in 1962 when a simple detector was boosted by a rocket above Earth's atmosphere for a few brief minutes. When you consider that the increase in capability of Chandra over this detector is comparable to the increase in capability from Galileo's telescope to the Hubble Space Telescope, you begin to sense the size and scope of this accomplishment. It is as if the equivalent of 400 years of technical development had been compressed into 37 years. That is fast work, even for fast times.

I

THE DREAM OF A LARGE X-RAY OBSERVATORY

1

High-Energy Vision

RICCARDO GIACCONI was only 27 years old, but he was already worried about his future. What had started out as a promising field of research was losing its luster. Four years earlier, in 1954, he had received a Ph.D. in physics from the University of Milan, working with Guiseppe Occhialini, one of the world's premier cosmic ray physicists. Giacconi had impressed his mentor with his ability to design and build sophisticated detectors. He was awarded a Fulbright fellowship to work at Indiana University, and then moved to Princeton, with the expectation that he would follow in Occhialini's footsteps and make some great discoveries.

In cosmic ray physics, a great discovery had traditionally meant the discovery of a new type of particle. Cosmic rays are high-energy charged particles that bombard the top of Earth's atmosphere. They are the nuclei of atoms, mostly hydrogen atoms, that have been accelerated to speeds close to the speed of light. Exactly how this acceleration occurs remains a subject of lively controversy in astrophysics to this day. But to elementary particle physicists in the first half of this century, cosmic rays presented not so much a problem as an opportunity. Their energies were much greater than could be attained in existing particle accelerators. Cosmic rays were the only way physicists could study high-energy nuclear reactions that would lead to the production and discovery of new subnuclear particles.

The search for the ultimate constituents of matter has always been at the forefront of physics, and in the 1930s and 1940s many of the important advances in the field were made by cosmic ray physicists. Besides their high energy, cosmic rays have another attractive feature—they are free, so countries such as England and Italy with a large supply of skilled physicists and a small supply of money can make important contributions. Positrons, the antimatter equivalent of electrons, were discovered by placing a suitably prepared target in the path of incoming cosmic rays, as were pi and mu mesons. Occhialini was involved in both those discoveries, and Giacconi hoped to make similar contributions.

The problem with cosmic ray physics is that it is really cosmic ray astronomy. You can't produce the cosmic rays on demand. In the words of the Nobel laureate physicist Emilio Segre, "One could observe them only when God sent them." Worse still, cosmic rays come from all directions, so you can't improve your odds by looking in a particular direction, other than up.

There are only two things cosmic ray physicists (they resist being called astronomers) can do to increase their chance of success. They can get above as much of the atmosphere as possible, to reduce the effect of absorption by the nuclei of the atoms in the atmosphere. And they can build as large and efficient a detector as possible.

While a graduate student, Giacconi had built the largest multiplate cloud chamber in Italy and had installed it at the Testa Grigia Laboratory, located 11,000 feet above sea level at the foot of the Matterhorn. In 2 years, he observed only 80 cosmic ray interactions and found no new particles. Although he enjoyed building new detectors with his colleagues Herbert Gursky and Fred Handel at Princeton, he was not satisfied with the scientific return on this work.

Like the person who came up with a better buggy-whip at the same time as the automobile was invented, Giacconi had a problem not with his talent but with his timing. By the late 1950s, powerful particle accelerators had been developed that could produce large fluxes of high-energy particles on command. It was clear that future major discoveries in high-energy nuclear physics would be made by the large groups affiliated with the new synchro-cyclotron accelerators at Berkeley, the University of Illinois, Brookhaven in New York, Dubna in the Soviet Union, and Geneva, Switzerland. Giacconi wasn't eager to join one of these large groups and be pushed into yet another long apprenticeship. He had to make a move. His visitor's visa was about to expire and he had no permanent position in Italy. He had recently married and had a family to support. His prospects were bleak.

Meanwhile, in Cambridge, Massachusetts, another young cosmic ray physicist had already made his move. Martin Annis, a recent graduate of MIT, was a new breed of scientist who sought not fame but fortune by exploiting a niche created by the Cold War and the space race. He rented part of a converted milk truck garage on Carleton Street in the shadow of MIT and set up American Science & Engineering (AS&E). He persuaded Bruno Rossi, his former professor at MIT and perhaps the most respected cosmic ray physicist in the world, to become chairman of the board. Then he set about recruiting bright, energetic young scientists who could get government contracts, presumably from the Defense Department and possibly from a new governmental organization that had just begun operation, the National Aeronautics and Space Administration (NASA). His principal source of talent and contacts was MIT. Another MIT professor and cosmic ray physicist, Herbert Bridge, had been impressed with a young physicist from the Princeton cosmic ray group whom he had met in the course of a joint experiment. He introduced Annis to Riccardo Giacconi. Annis immediately offered Giacconi a salary of $13,000 a year to start a program in space science at AS&E.

Giacconi arrived at AS&E in early 1959 with no particular plan in mind. The obvious choices of looking at charged particles in the radiation belts of the upper atmosphere, or building detectors to monitor nuclear bomb explosions in space, didn't appeal to him. He still wanted to do serious, fundamental research, but he knew little about his new field, space science. In fact, in those days hardly anybody knew what the field promised, which is why in 1958 the National Academy of Sciences had established a Space Science Board composed of eminent scientists, including Bruno Rossi, to help NASA formulate a space research strategy.

As a member of the Space Science Board, Bruno Rossi had heard various subcommittees discuss a number of interesting suggestions, but one had especially intrigued him. John Simpson of the University of Chicago, Leo Goldberg of Harvard University, and Laurence Aller of the University of California had recommended a survey of the sky with X-ray detectors. In one important respect it was an ideal project for space research. The atmosphere of Earth would absorb cosmic X-rays if they existed, so the only option available to would-be X-ray astronomers was to put detectors 100 miles or more above the Earth.

Herbert Friedman and his colleagues at the Naval Research Laboratory (NRL) in Washington, D.C., had been trying to detect cosmic X-rays since the late 1940s with detectors aboard V-2 and Aerobee rockets. They had established that the sun is a source of X-rays, but had been unable to detect X-rays from other stars or any other cosmic source outside our solar system. The reason was clear to most astronomers: The amount of X-radiation produced by the sun is small, a million times less than its optical radiation. If the radiation from the sun is typical, then detectors millions of times better than those used by Friedman's group would be needed to detect X-rays from other stars. The majority of the Space Science Board soon forgot about the proposed X-ray survey, because too little was known.

For precisely the same reason, Rossi couldn't let the idea go. His experience during the glory years of cosmic ray research had instilled what he described as "a deep-seated faith in the boundless resourcefulness of nature, which so often leaves the most daring imagination of man far behind." Nor did he underestimate the difficulties. Major state-of-the-art improvements of existing detectors would be required and space research involved high inherent risks, not the least of which was that the rocket could blow up on the launch pad. He could help, but he knew a younger scientist would be needed to carry the torch. He also knew, as chairman of the board of AS&E, that they had recently hired a young physicist, who came highly recommended from his old colleague and compatriot Occhialini, to develop a space science program at the company. He invited Giacconi to a party at his home. Within minutes after their first meeting, Rossi made his proposal: Giacconi should use his position as director of space research at AS&E to develop a program in X-ray astronomy.

For Giacconi, it was a revelation, a bolt from the blue. "All of a sudden I had a way to go," he recalled. And go he did. For the next 10 years, Giacconi worked, by his own account, "as a man possessed" to open up a new window on the universe.

2

Invisible Light

INVISIBLE LIGHT. The phrase seems to be a contradiction in terms. We tend to think of light as intimately related to the experience of seeing. Every day at dawn the first rays of sunlight make visible a world rich in color and detail. Or the flip of a light switch allows us to see even though it is dark outside. Darkness is the absence of light. We cannot see objects or colors in the dark. How then can light be invisible?

In 1799, William Herschel, who became famous for his discovery of the planet Uranus, was studying sunlight. Herschel was the premier astronomer of his time. With his sister Caroline, he built the best telescopes in the world, and he used them to make fundamental contributions to virtually every area of astronomy—planets and their moons, the sun, double stars, star clusters, and nebulae. He also had some bizarre ideas, chief among them that the interior of the sun was a cold solid body that might even be inhabited. Sunspots, in this view, could be holes in the sun's fiery atmosphere through which the cooler surface could be seen. Herschel trained some of his most powerful telescopes on sunspots to examine them in as much detail as possible.

These investigations were hampered by the intolerable heat produced by focusing the sun's light. Isaac Newton had shown over 100 years earlier that sunlight breaks up into a rainbow, or spectrum, of colors when passed through a prism. Herschel experimented with differently colored glasses placed behind the telescope's eyepiece to filter out the different colors of sunlight, and he undertook a series of investigations to determine in which color the sun was brightest, and which color produced the most heat. He found that the spectrum of sunlight reached a maximum intensity between yellow and green. He then allowed the light from the sun to fall on a thermometer, one color at a time, to see if there was any difference in the amount of heat produced by the different colors of sunlight. The result astonished him.

The temperature rise in the thermometer was highest at a location beyond the red end of the spectrum, where no light could be seen! After checking this result repeatedly, he boldly concluded that sunlight consisted in part "if I may be permitted the expression, of invisible light." In over 200 subsequent experiments, Herschel showed that infrared light, as it came to be called, was produced by fires and other heated objects, and obeyed the same laws of reflection and refraction as visible light.

Herschel's thoroughness and ability to focus on a problem led him to many important discoveries. His contemporary William Wollaston, also a brilliant scientist, had quite a different style. His constantly shifting focus led to many discoveries, but his failure to follow through caused him to miss some even greater ones. Wollaston developed a method for molding platinum that made him independently wealthy, discovered the elements palladium, rhodium, and titanium and the amino acid building block cystine, and made contributions to mineralogy, optics, acoustics, physiology, botany, and astronomy. He also misidentified niobium as tantalum, thus precipitating a debate that lasted for a generation. He was the first to note the presence of dark lines in the spectrum of sunlight, but failed to follow up on the extraordinary implications of this discovery.

In another missed opportunity, Wollaston was one of the first to observe ultraviolet light, but the credit went to the German physicist Johann Ritter, whose research was more thorough. In 1801 Ritter found that the invisible region of the spectrum beyond the violet contained energetic rays which were more effective than visible light in breaking down silver chloride, a key chemical reaction involved in photography.

Almost two decades later, Wollaston showed up again with a chance to play a leading role in the story of light, and once again he wound up as a supporting actor. This episode had to do with the relation between electricity and magnetism, which, as we shall see, is fundamental to understanding the nature of light.

In 1819, Hans Christian Oersted, a Danish physicist, had created a sensation among scientists in Europe and the United States with his discovery that an electric current produces a magnetic force. Oersted showed that if you bring a compass near an electric current, the compass needle, which is a small magnet, will be deflected by the magnetic force of the current.

Wollaston, with the keen instincts of a successful inventor, realized the importance of this discovery. What if the deflection could be turned into a rotation? And what if the reverse were true, that a rotating magnet could drive an electric current? All sorts of practical applications—electric motors, electrical generators—could be imagined. Remember, this was a time when electrical currents could be produced only with batteries, unless you were foolhardy enough to try to duplicate Ben Franklin's experiments with lightning.

Wollaston's intuition was right on target, but his experimental skills were not up to the challenge. Nor could his formidable colleague Humphry Davy crack the problem. Davy had developed giant batteries and used them to pioneer the field of electrolysis, a technique for separating elements in compounds by running large electric currents through solutions. He was the first to isolate potassium, sodium, barium, strontium, calcium, magnesium, and chlorine. But he was unable to demonstrate that the electric current could produce continuous motion, or whether a rotating magnet could drive an electric current.

Wollaston was a good scientist and Davy was a great one, but Michael Faraday, Davy's 30-year-old assistant, a self-educated bookbinder turned physicist, was a genius. After hearing of Wollaston and Davy's efforts, Faraday constructed an elegant but simple apparatus consisting of a battery, two beakers full of mercury (to complete the electrical circuit), hinged wires, and moving magnets, and showed that electrical and magnetic forces could be converted into continuous mechanical movement. He had invented the electric motor.

For this and other work, Faraday was nominated for election to the Royal Society of London. Davy, his mentor, opposed the nomination on the grounds that Faraday had gotten his idea of the experiment from Wollaston and himself. Wollaston, however, took Faraday's side and Faraday was elected with but one negative vote on the customary secret ballot.

The invention of the electric motor was significant, but it would have limited practical value unless Faraday could close the loop on Oersted's discovery. If he could use rotating magnets to drive a current, then it would be possible to generate electricity on a grand scale from readily available forms of energy such as falling water or heat from burning wood or coal. More important to Faraday, who was not particularly interested in practical applications or fame or fortune, it would confirm his conviction that electricity and magnetism "have one common origin; or in other words are so directly related and mutually dependent that they are convertible, as it were, one into another."

In 1831, Faraday and, independently, Joseph Henry, an American mathematics and philosophy teacher at Albany Academy in Albany, New York, showed how electricity can be produced from a moving magnet. Henry did not follow up on his studies, but Faraday worked tirelessly until he had constructed the first electric generator—a disk of copper rotating between the poles of a magnet—that would produce a continuous flow of electric current. Electricity and magnetism are inseparably linked, like the chicken and the egg. Each can be induced from the other. This discovery of electromagnetic induction is the basis for electromagnets, electrical generators, and a thousand other devices that make modern life possible.

What struck Faraday was that the conversion of electrical energy into magnetic energy and vice versa was accomplished without transferring matter. He developed a theoretical model according to which magnets and electric currents produced lines of force that spread out into space around the magnets or currents. This force field would then act on a magnetic object, such as a compass needle, and cause it to line up with the field. In this picture, Faraday's discovery of electromagnetic induction had shown that a changing magnetic field could create an electric force field and a changing electric force field could create a magnetic force field. One could imagine a disturbance in an electric field propagating across empty space and inducing a disturbance in the magnetic field, which in turn would produce a disturbance in the electric field that would lead to a measurable effect at some distance. And if one were Michael Faraday one could imagine that all this might have something to do with another mystery, the nature of light.

By 1841 Faraday was mentally exhausted by a decade of intense research. He suffered a mental breakdown, complained of memory loss, and retreated into semi-retirement. Then in 1845 he received a letter from

William Thomson, a 21-year-old prodigy who would go on to a long and illustrious career and become known as Lord Kelvin. Thomson was trying to work out a mathematical description of Faraday's fields, and wondered if there might be a relation between electric and magnetic fields and light. This idea appealed to Faraday, who believed in the underlying unity of all physical phenomena, so he set to work to prove the connection. Faraday was able to demonstrate that magnetic fields affect the properties of light, but he could not prove what the connection was. Nor could Thomson prove it mathematically. That accomplishment was left to a young Scottish mathematician and physicist whose genius was very different from, but equal to that of Faraday's.

3

Light Quanta

JAMES CLERK MAXWELL came from a world very different from Faraday's. Maxwell was well born, well educated (second in his class at Cambridge), and a professor by the time he was 24. He was a great admirer of Faraday's work. Whereas many physicists had ignored Faraday's theories about electric and magnetic fields because of his lack of mathematical sophistication, Maxwell was profoundly impressed by Faraday's careful research and physical intuition. He suggested that Faraday might have been stifled if he had been "a professed mathematician" because "he did not feel called upon either to force his results into a shape acceptable to the mathematical taste of the time or to express them in a form which mathematicians might attack." Such a suggestion would not be well received by today's physicists, nor was it in Maxwell's time. Nevertheless, Maxwell did not equivocate in his advocacy of Faraday's insights. In his monumental *Treatise on Electricity and Magnetism* he wrote, "It is mainly with the hope of making these [Faraday's] ideas the basis of a mathematical method that I have undertaken this treatise."

How well Maxwell succeeded can be judged from the comments of Richard Feynman, a modern-day genius who contributed much to our understanding of the nature of light, that "there can be little doubt that the most significant event of the 19th century will be judged as Maxwell's discovery of the laws of electrodynamics." Albert Einstein and his collaborator Leopold Infeld called Maxwell's discovery of the equations describing electromagnetic phenomena "the most important event in physics since Newton's time." What Maxwell did was work out a set of four equations that explained all of Faraday's work and more. Maxwell's equations showed how electric and magnetic fields are produced by electric charges and currents, how magnetic fields are induced by changing electric fields (electric motors, electromagnets), and how electric fields are induced by changing magnetic fields (transformers, generators, videocassettes, computer disks).

The equations also implied that a wave of rapidly changing induced electric and magnetic fields would move out into the space surrounding a changing electric current. The current would induce a changing magnetic field that would induce a changing electric field that would induce a changing magnetic field, and so on. Maxwell found that this cycle could be self-sustaining, that is, the wave could go on forever and travel vast distances, if a certain condition was satisfied—that the waves travel at a precise speed. That speed was 186,000 miles per second, the speed of light.

Now he could add another monument to his edifice of discovery: that light is an electromagnetic wave produced by changing electric currents.

The stunning implications of this discovery were immediately obvious to Maxwell. If light is produced by changing electric currents, which are moving electric charges, then it should take many forms, depending on how rapidly the electric charges are accelerated. For example, imagine yourself in a swimming pool, or a very large bathtub, or a still forest pond, whichever you prefer. If you move your hand back and forth in the water, you will create a wave that spreads out across the pool. If you move your hand more rapidly, the waves will move out with a higher frequency. The distance between the crests of the waves, called the wavelength, will be smaller.

Electromagnetic waves are like water waves in that the frequency of the wave depends on how rapidly the electric charges producing the wave are oscillating. Visible light represents one special range of light corresponding to one specific range of charge oscillations, between about 500 trillion and 1,000 trillion oscillations per second. Infrared light represents a lower frequency of oscillation than red light, which in turn has a lower frequency than blue light, which in turn has a lower frequency than ultraviolet light. In principle, Maxwell predicted, the frequency of light waves that could exist depends only on how slowly or how rapidly electric charges can oscillate. He knew that other types of light waves should extend far below the infrared and far above the ultraviolet.

Another consequence of Maxwell's insight into the nature of light is the understanding that one of the things we know and love most about visible light—its color—is not a fundamental property of light at all. A prism spreads out the different wavelengths of visible light. We perceive these different wavelengths as various colors. The perception of color is probably an evolutionary adaptation to help us pick up more detail in an environment filled with light in the visible wavelength range. A person with defective color vision and many animals may not agree that a rose is red, but they would have to agree that its wavelength is about 20 millionths of an inch.

Maxwell died in 1879 at the age of 48, so he did not live to see his bold prediction of the existence of other types of light waves proved by the German physicist Heinrich Hertz in 1888. Hertz constructed an electric circuit that produced a rapidly oscillating electric spark and showed that it generated electromagnetic waves. The wavelength of these waves was about 60 centimeters, corresponding to a frequency of about 500 million oscillations per second, or 500 megahertz as we call it today in honor of Hertz. A few years later, when Guglielmo Marconi, an Italian engineer, learned of Hertz's work he recognized its practical implications and set to work developing a means of using electromagnetic radiation for wireless telegraphy, called radiotelegraphy or radio.

During the time when Marconi was developing the radio, an extremely high frequency type of electromagnetic radiation was discovered. In November of 1895 Wilhelm Roentgen, a German physicist, discovered a highly penetrating form of radiation that he called X-radiation to emphasize its unknown nature. Roentgen suspected that X-rays were a form of electromagnetic radiation, but was unable to prove it, because he could not reflect the waves. Nor could he produce a wave interference pattern in which the peak of one wave coincides with the trough of another, producing a cancellation of the wave. This does happen, for example when light is reflected off a soap bubble or an oily spot on a wet street. Waves that reflect off the top of the film of a soap bubble can interfere with waves that reflect off the bottom of the film. Depending on the thickness of the film and the angle of reflection, you see a spectrum of colors as different wavelengths are subtracted from the incoming light, which is usually a mixture of colors.

Max von Laue, another German physicist, came up with an explanation as to why Roentgen had failed to observe reflection or interference of X-rays. The frequency of X-rays is so high, and the wavelength is so small, comparable to the size of an atom, that they would reflect only if the angle of reflection were very small and the reflecting surface were very smooth, and they would interfere only if films with the thickness of a few atoms were used. Von Laue reasoned that the layers of atoms in a crystal would be closely enough spaced to produce an interference pattern with X-rays. In 1912, von Laue's colleagues Walther Friedrich and Paul Knipping produced an X-ray interference pattern in the laboratory by exposing a crystal of copper sulfate to a beam of X-rays for several hours. This established that X-rays are a high-frequency, short-wavelength form of light. During this period an even higher frequency form of light, called gamma rays, was discovered by researchers studying radioactive elements.

By 1920, the electromagnetic spectrum from low-frequency radio waves to high-frequency gamma rays was under active investigation. A new theory—quantum theory—governing the behavior of matter at the atomic level was also being put together. The cornerstone of this theory is the revolutionary idea, first put forward by the German physicist Max Planck in 1900, that the energy states of the fundamental particles that constitute the building blocks of matter are not continuous, but discrete. In other words, the energy of one of these particles cannot be increased or decreased by an arbitrarily small amount. Rather, it must be increased or decreased by a discrete amount, called a quantum of energy. For example, a quantum of energy could be likened to stepping up or down stairs. You can step one stair at a time or two or three stairs at a time, but you cannot step one and a half or three and a half stairs at a time.

In 1905 Einstein proposed a bold extension of Planck's idea. He argued that the quantum hypothesis applied to light as well as particles, that light waves are not emitted continuously, but in quanta: bundles of pure energy called photons that travel at the speed of light. A stream of many photons behaves like the electromagnetic waves described by Maxwell. The energy of the photons determines the frequency and wavelength of the wave, with high energy corresponding to high frequency or short wavelength.

As Emilio Segre said, "At that time scientists knew that light was made of electromagnetic waves; if anything was certain, that was it." Einstein's paper was a real shocker to his colleagues. Not that Einstein had any col-

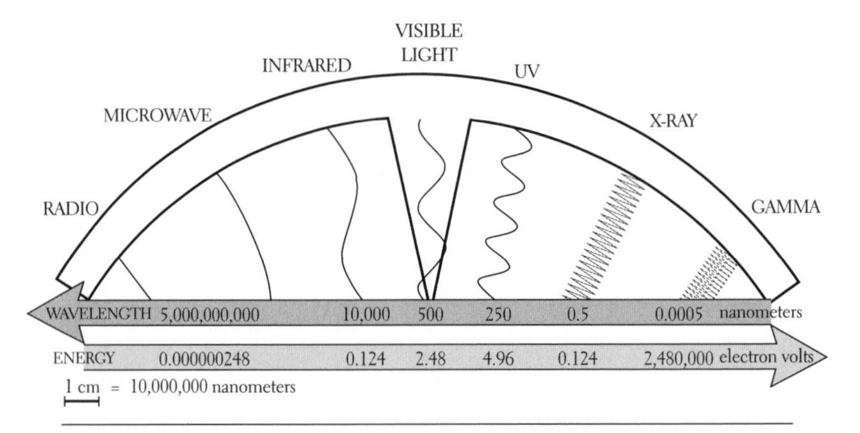

The electromagnetic spectrum. (SAO/CXC/S. Lee).

leagues in the real sense of the word, except possibly for his friend the mathematician Marcel Grossman. Then, none of the members of the physics establishment knew who he was. Yet his reasoning was clear and compelling and he gave several examples of puzzling observations that his hypothesis would explain. And there was that other paper he had published in the same year on "the principle of relativity" with its intriguing notions about space and time.

Within a few years virtually every physicist knew who Albert Einstein was. He had his pick of professorships at all the important universities and was nominated for admission to the elite Prussian Academy of Sciences. Yet the idea of photons still stuck in the craw of the old guard, who acknowledged Einstein's many brilliant contributions to a number of problems, but thought that he missed the target in his hypothesis of light quanta.

X-rays played a major role in the confirmation and development of quantum theory due to the property that made them so unlike visible light, namely the high frequency or, equivalently, the high energy of X-ray photons. The energy of an X-ray photon is high enough to tear the electrons orbiting near the nucleus of an atom away from the atom in a process called photoelectric absorption. Physicists can use X-rays to dismantle the atom and study how the electrons are arranged in their orbits around the nucleus of the atom. In this manner, it was determined that the atoms of each element have a unique architecture which restricts the electron's motions to specific orbits.

The same property that makes X-rays so useful for studying atoms makes X-ray telescopes impossible to use from the ground. Cosmic X-rays interact strongly with the atoms in the Earth's atmosphere. Even though the atoms in the atmosphere are widely spaced, the total thickness of the atmosphere is large and the total number of atoms is enormous. The chance that an X-ray produced in outer space can make it through Earth's atmosphere without colliding with an atom and being absorbed is like the chance of driving through Los Angeles during rush hour without having to slow down and stop—nil.

At first, the only way to observe X-rays from the cosmos was to put a detector high above the Earth on a balloon (this works for only the highest-energy X-rays), rocket, or satellite. Adding to the adversities of the would-be X-ray astronomer was the problem that Roentgen encountered: Because of their high energies, X-rays are extremely difficult to focus. If you can't focus them, you can't build a true X-ray telescope, and if you can't build a telescope, you will always view the X-ray universe "through a glass darkly."

4

The Birth of X-ray Astronomy

RICCARDO GIACCONI'S incisive intellect cut quickly to the heart of the matter. He began searching the literature for clues as to how to build an X-ray telescope. A review article by the German physicist Siegfried Flugge described the physics of X-ray reflection. In essence it is this: because of their high energy or short wavelength, X-rays will reflect off a mirror only if they strike it at a grazing angle of a few degrees or less. (A degree is approximately the angle subtended by your thumb held sideways at arm's length.) Think of skipping a pebble on a pond.

Another German physicist, Hans Wolter, had published a paper in 1952 in which he showed that not one, but two reflections were needed to produce a two-dimensional image with grazing incidence reflection. Wolter had in mind building an X-ray microscope, but he quickly realized that the short wavelength of X-rays would defeat his efforts. The surface of the mirror would have to be smooth to within a tolerance of a few tens of atoms. Otherwise, the incoming X-rays would plow into the bumps on the mirror like a skipping pebble encountering a wave rippling across the surface of the pond. Grinding and polishing the mirrors for a microscope to the appropriate shape and smoothness would not be practical. Wolter dropped his idea, but Giacconi picked it up 7 years later. Since the mirrors for a telescope would be much larger than for a microscope, the technical problem of grinding them to the proper shape should not be insurmountable, Giacconi figured.

Over the next 10 years, he would learn that not everyone agreed with this optimistic view, especially those in control of the funds. Bruno Rossi was impressed, though, and for the time being that was all that mattered

Schematic of a grazing incidence X-ray telescope. (SAO/CXC/D. Berry.)

since he was chairman of the board and a recognized expert in optics. Within months, Giacconi and Rossi published a paper on the design of X-ray telescopes, some 2 years before any sources of cosmic X-rays other than the sun were known to exist.

The reflective properties of X-rays dictate a design that is radically different from mirrors for optical telescopes. Remember, mirror surfaces for an X-ray telescope must be almost parallel to incoming X-rays. The design for a single set of mirrors that could make an image consisted of two cylindrical glass barrels (one for each of the two required reflections) with a slight, precisely defined taper from the front to the back. The second glass barrel was slightly smaller in diameter. Imagine two large drinking glasses with the bottoms cut off, end to end. One difficulty with this arrangement was immediately apparent. The target area, or the effective area, to use the astronomers' term, was small because only a small range of angles for the incoming X-rays will result in a reflection. This reduced the effective area to 5 percent or less of the total area of the mirror. Giacconi and Rossi suggested that one way to compensate for this lack of effective area was to nest smaller cylinders inside larger ones like Russian dolls.

Giacconi began work on a prototype X-ray telescope immediately. At the same time, he realized that his efforts would be futile unless it could be shown that many cosmic X-ray sources existed. This would have to be demonstrated with conventional detectors, which were in essence small boxes about the size of pill boxes. The boxes had a very thin window made of mica to let X-rays through and were filled with a gas such as argon that absorbs X-rays efficiently. Giacconi submitted a proposal to NASA. It was rejected. Undeterred, he turned to the U.S. Air Force, which through its Air Force Cambridge Research Laboratories (AFCRL) was already funding classified research at AS&E on the detection of high-altitude nuclear explosions.

The air force was interested in studying solar flares for the effect they might have on the disruption of radio communication. Using the fact that Herbert Friedman and his colleagues at the Naval Research Laboratory had discovered that active regions of the sun emitted X-rays, Giacconi's group made a bold but successful proposal to AFCRL: they would study X-rays from the sun *and* the moon.

Their reasoning was that high-energy particles streaming away from the sun would strike the moon and produce a detectable flux of X-rays. This clever proposal made it possible to go for much bigger game. By observing in a direction away from the sun, they were opening the door to the detection of an X-ray source from beyond the solar system, if nature would be so kind.

On the first try, the Nike-Asp rocket engine failed. AFCRL gave them another chance 16 months later. Meanwhile Giacconi continued to build a dedicated team of scientists and engineers, which now included Frank Paolini, an expert at building these types of detectors, and Herb Gursky, a cosmic ray physicist and colleague from Giacconi's Princeton days. Using techniques borrowed from cosmic ray physics, they developed a detector that was 100 times better than any previously flown. The second rocket engine worked, but the doors in the sides of the rocket did not open, so all their new sensitive detector saw was the inside of the rocket door.

The third try was a charm, though, and 1 minute before midnight on June 18, 1962, the launch occurred that opened the exploration of the high-energy universe. For 350 seconds, X-ray photons trickled in from the depths of the cosmos at the rate of a few per second. When the rocket window spun toward the constellation Scorpio, the count rate jumped dramatically. Giacconi's group had discovered the first X-ray source outside the solar system. Sco X-1, as it came to be called, surpassed all hopes and expectations in its brightness as a cosmic X-ray source.

In retrospect, the failure of the first rocket had been a marvelous piece of good luck for Giacconi's group. If that rocket had worked, the detectors on board would not have been sensitive enough to detect Sco X-1, and it might have been years before they could have obtained funding to try again. In that case, they most likely would have been scooped by Friedman's group at the Naval Research Laboratory, who demonstrated their readiness by providing confirmation in April 1963 of the existence of Sco X-1 and discovering another strong X-ray source associated with the remnant of an exploded star.

What about the search for X-rays from the moon? They were not detected until 28 years later, when the Roentgensatellite (Rosat), a European X-ray observatory, detected some faint X-ray fluxes from the moon.

Giacconi moved quickly to press his advantage with NASA, which had missed a bet by refusing to fund AS&E's history-making rocket flight. With only two cosmic X-ray sources discovered, a cautious science administrator might have been excused for wanting to wait to see what developed, to test the waters with a toe or perhaps even a foot. Nevertheless, Giacconi presented another option to Nancy Roman, the chief of the astronomy branch of NASA: dive in headfirst. Undertake immediately a long-range program for X-ray astronomy.

Begin by funding an expanded program of rocket flights to survey the sky and determine just how common cosmic X-ray sources are. Follow up quickly with an instrument on the Orbiting Solar Observatory series of spacecraft, then an Explorer satellite dedicated to X-ray detections, then a small focusing X-ray telescope, and finally a large focusing X-ray telescope, 30 feet long, that could provide X-ray images on a par with optical images.

Roman refused to dive in, but she did choose to wade decisively into the unexplored waters of X-ray astronomy. She considered X-ray telescopes premature, but endorsed the concept of a satellite dedicated to X-ray observations. Jubilant, Giacconi and his group submitted an expanded proposal for an X-ray satellite with a suggested launch date of December 1965. Their proposal was approved, but they were soon to come up against several reality checks concerning how business and science are done with NASA.

First, a small company such as AS&E could be entrusted with the construction of the scientific payload, but not with the construction of the spacecraft. Second, every NASA space science project is managed through one of its field centers, such as the Goddard Space Flight Center in Greenbelt, Maryland, or the Jet Propulsion Laboratory in Pasadena, California, or the Marshall Space Flight Center (MSFC) in Huntsville, Alabama. It is not enough that one of these centers agrees to accept management responsibility for a program. Someone high up in the organization has to be enthusiastic about it, or it will never survive the endless, debilitating struggles necessary to acquire funds and keep them flowing in to support the program.

In those days a small, innovative project such as the X-ray Explorer stood little chance of getting through the NASA bureaucracy. But a large program—there is a profound distinction of size and importance in the minds of NASA officials between a project and a program—might find a supporter with enough clout to make it happen. This takes time, which is the third point on the learning curve of working with NASA. Nothing happens as quickly as you think it will happen.

Giacconi found his highly placed advocate in John Naugle, who was associate administrator for science and applications at NASA. Two and a half years after the submission of the X-ray Explorer proposal and a year after the proposed launch date, NASA announced that it would begin a program of Small Astronomy Satellites. The X-ray Explorer would be the first and the Goddard Space Flight Center would manage the program. The launch date was set for late 1970.

Meanwhile, Giacconi continued to pursue his dream of an orbiting observatory with a large focusing X-ray telescope. He formed a collaboration with John Lindsay at the Goddard Space Flight Center, who used his influence to secure funding from NASA's solar physics branch for the development of an X-ray telescope to study the sun. They produced an X-ray telescope about the size of a mailing tube that made telescopic X-ray images of the sun from sounding rockets for the first time in October of 1963 and again in March of 1965. NASA was impressed and plans to put a much larger solar X-ray telescope on a manned spacecraft were under way when Lindsay, an apparently healthy man in his forties, suffered a fatal heart attack while mowing his lawn.

The AS&E-Goddard collaboration was dissolved, but Lindsay had made his mark, both through his technical expertise and through his

Early X-ray telescopes. (L. van Speybroeck.)

advocacy within NASA. Research and development on solar X-ray telescopes continued at both AS&E and Goddard.

Despite the successes of the early solar X-ray telescope flights, NASA officials and scientific opinion-makers saw no urgent need for an X-ray telescope to study X-ray sources outside the solar system. They preferred to wait and see if all-sky surveys with conventional instruments would show that an X-ray telescope was really needed. What if X-ray astronomy turned out to be the study of a few peculiar stars?

Giacconi argued the case of an X-ray telescope to whomever would listen at whatever forum, in personal conversations, at lectures, at presentations before scientific advisory committees. At one auspicious scientific advisory committee meeting in Woods Hole, Massachusetts, in the summer of 1965, the X-ray and Gamma Ray Astronomy Panel of the Space Science Board, chaired by Herbert Friedman, convened to study plans for future X-ray and gamma ray observations.

Frank McDonald, an influential cosmic ray physicist from Goddard Space Flight Center, was there, pushing an intriguing idea. The Apollo lunar exploration program was in high gear and was producing lots of large spacecraft, which were relatively inexpensive. Why not start a program—remember, with NASA you have to think big to get anyone's attention—of cosmic ray, X-ray, and gamma ray satellites that would use identical spacecraft? Giacconi pressed for inclusion of an X-ray telescope in McDonald's proposed program as a natural and essential follow-up to a large survey-type experiment proposed by Friedman. The panel agreed, as did NASA's Astronomy Missions Board 2 years later. Feasibility studies of a high-energy astronomy observatory began in 1968 at the Marshall Space Flight Center.

Meanwhile, the field of X-ray astronomy was progressing rapidly. Using balloons and rockets to launch their detectors, groups from AS&E, Rice University, the University of California at San Diego, Lawrence Livermore Laboratories, Lockheed—where Philip Fisher, an ingenious experimenter, was developing his own version of an X-ray telescope—the Naval Research Laboratory, MIT, and the University of Leicester had discovered more than 30 sources of X-rays. About half a dozen sources were associated with supernova remnants—the remnants of exploded stars—and a few were associated with galaxies. The remainder were bizarre, X-ray–emitting stars whose exact nature was a mystery. Technical advances were also occurring apace in an environment that encouraged an open exchange of ideas concerning the instrumentation as well as the science.

The scientists and engineers at AS&E studied each new development carefully with an eye to how it could improve the design of the X-ray Explorer. The unusual nature of AS&E in the 1960s helped this process. As a private company that was expanding rapidly into pure scientific research as well as commercial applications and defense research, it could compete favorably in the job market by offering good salaries and the opportunity to focus on research without the distractions of teaching, faculty meetings, and other academic duties. In fact, the scientists and engineers had to focus on their work, because there was no tenure system. A lack of competence or dedication could be quickly identified and even more quickly remedied. The absence of a safety net was no deterrent, especially to the young scientists, because under Giacconi's leadership AS&E was increasingly becoming known as a place where the action was in space astronomy.

Gursky joined the group in 1960. A few years later, Paul Gorenstein, a young nuclear physicist from MIT and a Fulbright fellow, added his considerable experimental talent to the team. Leon van Speybroeck, another new Ph.D. from MIT, came shortly after Gorenstein and began working on X-ray mirrors. In 1968, Harvey Tananbaum, who had just completed the first Ph.D. thesis from MIT in the new field of X-ray astronomy, came to work at AS&E on the X-ray Explorer program under the direction of Ed Kellogg, who had received his Ph.D. from the University of Pennsylvania a few years earlier. I (WT) had written a Ph.D. thesis on X-ray astronomy at the University of California at San Diego. My thesis involved theory rather than observation, so I had little thought of joining the migration to AS&E. Yet when Gursky called to offer me the opportunity to set up a small theoretical physics group in the spring of 1969, the temptation to be there when the X-ray Explorer flew was too great. Against all advice from my colleagues of the dangers of leaving a tenure-track job for the uncertain cut-throat corporate world, we packed up and moved to the Boston area.

By the late 1960s, AS&E had contracts in six different areas of space research and its defense-related research was being phased out. Giacconi was now on the board of directors and was executive vice president of the company. Most of the scientists at AS&E were younger than 30 years old and it was a time when "don't trust anyone over 30" was often heard in the news, at the coffeehouses, and on the streets. Most of the young scientists probably shared my diffidence about making the transition to the corporate world from academia, the only world we had known, presumably moving from a milieu dedicated to the pursuit of ideas to one dedicated to the pursuit of profits.

As it turned out, the word "profit" was scarcely ever mentioned. Instead there were frequent spirited discussions about what the X-ray Explorer would discover with its first prolonged view of the cosmos or about what a large X-ray telescope might image.

Even as the group rushed to prepare for the launch of the X-ray Explorer, Giacconi pushed relentlessly to keep the large X-ray telescope on track. A consortium of groups from AS&E, Columbia, the Goddard Space Flight Center, and MIT was formed and a proposal for a telescope with five pairs of mirrors, the largest with a diameter of 42 inches, was submitted in May of 1970.

The proposal contained theoretical predictions of what might be seen, but everyone realized that they were mostly daydreams. Until the X-ray Explorer flew, no one could know for sure what was out there. So far there were a number of teasing glimpses from 5-minute rocket flights of sources, some of which seemed to come and go. Were the mysterious X-ray stars related to pulsars, the rapidly spinning neutron stars that had recently been discovered by radio astronomers? Or were they another, unknown type of object? What about quasars, those bizarre objects that seemed to be improbably distant and incredibly bright—would they emit X-rays?

In November of 1970, the X-ray Explorer was shipped to Kenya in preparation for launch from the San Marco platform, an old oil-drilling rig modified by the Italian Space Agency for rocket launches. Why Kenya and not Cape Canaveral? A satellite launched near the equator gets a boost from the spin of the Earth, so the payload could be a little heavier. Another advantage of an equatorial orbit is that it avoids troublesome regions of the Earth's magnetic field where the concentration of charged particles is large, and therefore detector background noise is high. Harvey Tananbaum and two AS&E engineers, Dick Goddard and Stan Mickiewiz, accompanied the Explorer to Kenya as the AS&E component of the launch preparation team. They were joined by Giacconi a few weeks later. On December 12, 1970, Kenya's independence day, the X-ray Explorer, which was about the size and weight of a large television set, was launched into a 500-mile-high orbit by a Scout rocket. In appreciation of the cooperation of the Kenyan people, the X-ray Explorer was renamed Uhuru, the Swahili word for "freedom."

5

X-ray Stars

DURING THE 1960S scientists began to suspect that X-ray stars were related to the end-game of stellar evolution and they felt that there was something profound to be learned by studying them. These puzzling sources were producing a thousand times more energy in X-rays than visible light, in striking contrast to the sun, which radiates a million times more energy in visible light than in X-rays.

The sun and other normal stars shine because of nuclear reactions that convert hydrogen into helium deep in their interiors. For billions of years the outflow or energy from these reactions provides the pressure necessary to keep the star from collapsing under its own weight. After about 10 billion years, the supply of hydrogen in the core of a mid-sized star such as the sun will be exhausted. The star will then enter the stellar equivalent of a midlife crisis in which it tries to postpone the inevitable. Slow collapse produces nuclear fusion reactions in a shell of hydrogen on the edge of the collapsed core. This power surge expands the outer layers of the star a hundredfold to produce a bloated red star called a red giant. The sun is expected to turn into a red giant in about 5 billion years. When this happens, the mountains on Earth will melt and the oceans will boil away.

During the red giant phase, the outer layers of the star will evaporate like steam from a pot of boiling water. After about 100 million years the star will have emptied all its nuclear fuel tanks. With nothing to hold it up against the crush of gravity, the star will collapse to form a small dense sphere called a white dwarf—because it is white hot—and it will remain hot for a billion or so years from the heat generated in the collapse. When it becomes a white dwarf, all of the material in the star will be packed into a core about the size of the Earth, or about a millionth the star's original volume. This material is so dense that a piece the size of an olive would have the same mass as a mid-sized car.

White dwarfs presented a paradox to astronomers when they were discovered. What kept them from collapsing further? In 1926, R. H. Fowler, an English physicist, showed that the laws of quantum theory resolve the paradox. One of these laws is that no more than one electron can occupy any given energy state. To understand how this requirement makes white dwarfs possible, consider a parking tower outside a mall. Only one car is allowed per space. When business is slow, the parking lot is nearly empty, and there is very little movement among the cars. From time to time, a car will enter and park on the lower level, and one will leave, and that is it. As the holiday season approaches, however, the situation changes. As the parking lot fills, spaces are harder to come by. Cars race from one level to the next, as their drivers search for an empty space. The pressure builds.

The extremely dense matter of a white dwarf is like a parking lot at a mall on the last weekend before Christmas. All of the low-energy levels are degenerate, or filled, so the electrons are forced into higher-energy states not because they are being supplied with energy from another source, but because there is nowhere else to go. This degenerate electron pressure is sufficient to keep the white dwarf stars from collapsing under their own weight—up to a point.

In 1930, the astrophysicist Subrahmanyan Chandrasekhar showed that Fowler's solution held only for white dwarfs with a mass less than 1.4 times the mass of the sun. This limit became known as the Chandrasekhar limit. Collapsed stars more massive than the Chandrasekhar limit are doomed to continue collapsing beyond the white dwarf stage. Collapse to what? That was the question, and the answer was unclear as of the mid-1960s.

One possible answer, proposed by the astrophysicists Fritz Zwicky and Walter Baade of the California Institute of Technology in 1934, was that a new type of degenerate star, a neutron star, would result from the collapse of stars more massive than the Chandrasekhar limit. They had arrived at this conclusion in a roundabout way.

Baade and Zwicky had been investigating a class of eruptive stars called novas, in which previously undetected stars suddenly flare up to a luminosity that is 20,000 times that of the sun. They had come to the conclusion that another, more extreme class of exploding stars existed, which they called supernovas. Supernovas blaze forth in a matter of days with the brilliance of more than a billion suns, and a large fraction of the mass of the star is expelled into interstellar space at velocities of 20 million miles an hour. Baade and Zwicky speculated that such a large amount of energy could be released only if the core of the star collapsed to form the most compact star imaginable.

To see how collapse can generate energy, drop this book on a table. It is accelerated by the Earth's gravity and picks up energy of motion (kinetic energy), which is converted into sound energy or the energy of vibrations of the atoms in the table. To get a large amount of energy, from the collapse of a star, it must collapse as far as possible. This brings us back to the question—how far can a star possibly collapse?

Baade and Zwicky guessed that the answer was related to the structure of the atom and its nucleus. The atom nucleus is composed of positively charged protons and neutrons—neutrally charged particles that are a little more massive than protons. Circling the nucleus is a cloud of electrons. The diameter of the electron cloud is about a hundred thousand times the diameter of either the electrons or the neutrons and protons. This means that atoms, the basic building block of the solids, liquids, and gases that constitute our world, are mostly empty space.

Consider a rock. What we perceive as painfully solid when we bump into it is really the chaotic motion of electrons moving through empty space so fast that we cannot see or feel the emptiness. Suppose we generate enough pressure to squeeze all the empty space out a rock the size of a football stadium by crushing the electron clouds into the nuclei of the atoms. The 4-million-ton rock would be compressed to the size of a grain of sand!

What would happen if the mass of a collapsing star were greater than the Chandrasekhar limit? Baade and Zwicky speculated that it would squeeze the empty space out of the electron clouds, and force the electrons into the nucleus, where they would combine with the protons to form neutrons. Degenerate neutron pressure would finally stop the collapse, producing a neutron star with a mass about equal to that of the sun, and a diameter 1/100,000 that of the sun, or about 6 miles. The conversion of gravitational energy to heat in the collapse to form a neutron star would provide plenty of energy to explain a supernova explosion.

The theoretical properties of neutron stars were explored in the late 1930s by the Russian physicist Lev Landau and by Robert Oppenheimer and George Volkoff of the University of California at Berkeley. Oppenheimer then turned to the next problem. What would happen if the mass of the collapsing core of the star is so great that neutron degeneracy will not stop the collapse? The incredible conclusion reached by Oppenheimer and his student Hartland Snyder: the star would collapse indefinitely to form a gravitational warp in space.

Einstein's theory of General Relativity had shown that gravity curves space. His theory had been confirmed in one of the most dramatic observational tests in the history of science when the bending of light rays passing near the sun had been observed during a total solar eclipse in 1919. However, this was a bending of a fraction of a percent, a comparatively weak effect. The calculations of Oppenheimer and Snyder implied the strongest possible effect—the bending of light rays back in the direction from which they came. The gravitational fields around their hypothetical collapsed star were so intense, or the warping of space so great, that nothing, not even light, could escape. Everything was dragged inward by the implosion. In the words of Oppenheimer and Snyder, "the star thus tends to close itself off from any communication with a distant observer; only its gravitational field exists."

Any matter that came within a critical distance called the event horizon of the collapsed star would be pulled inexorably inward. There was no possibility of escape. The matter would be crushed without limit to an infinite density inside what John Wheeler of Princeton University so aptly named a black hole.

As discussed in detail by Kip Thorne in his popular book on black holes, these concepts "were too bizarre for most physicists in 1939." Surely something in nature would prevent such an extreme catastrophe, they argued. Perhaps the theory of General Relativity breaks down when the gravitational fields get extremely strong, or maybe the calculation of Oppenheimer and Snyder was too idealized. Shock waves or rotation or radiation must prevent the unending collapse. Something must prevent it.

The expertise and computing power necessary to test these ideas were not available until the 1960s. Then, calculations by Stirling Colgate and his colleagues at Lawrence Livermore Laboratories, and by Dave Arnett and Alistair Cameron, demonstrated that it was plausible, if not inevitable, that neutron stars were formed in the collapse of massive stars, and black holes were formed in the collapse of very massive stars. This work, together with the persistent and colorful advocacy of black holes by Wheeler and his colleagues, Thorne, and the Russian physicists Yakov Zeldovich and Igor Novikov made the concept of black holes worthy of serious consideration.

The general consensus in the early to mid-1960s concerning the ultimate fate of stars was that stars with a mass less than the Chandrasekhar limit of 1.4 times the mass of the sun would become white dwarfs. Everyone agreed on this. Most astrophysicists thought that a star with mass greater than the Chandrasekhar limit but less than about 5 times the mass of the sun would lose enough mass through stellar winds in its red giant phase so that it too would become a white dwarf.

What would happen to a star with a mass greater than 5 to 10 solar masses? Again, most astrophysicists believed that the collapse of such a star would produce a supernova explosion. After all, supernovas were well-studied phenomena by then, and they almost certainly represented the destruction of some type of star. Whether the explosion blew apart the entire star or left behind a neutron star was hotly debated. Unlike white dwarfs, which had been observed, neutron stars were only a theoretical construct; there was no observational proof of their existence. Scientists on either side could cite theoretical calculations that supported their positions, and point to obvious flaws in rival theories.

And then there was the possibility, contemplated by a growing minority, that the collapse of a massive star—just how massive was not at all clear, maybe more than 20 times the mass of the sun—would lead to the formation of a black hole.

All of the three theoretically possible end states of the evolution of a star—white dwarfs, neutron stars, and black holes—were considered candidates for explaining X-ray stars. Zeldovich and Novikov had discussed how matter falling onto a neutron star or a black hole would be accelerated to high energies. Shock waves near the surface of the neutron star or near the event horizon of a black hole would heat the infalling matter to millions of degrees and X-rays would result. It was as simple as falling off a log, or in this case falling off one star onto another. All you needed was

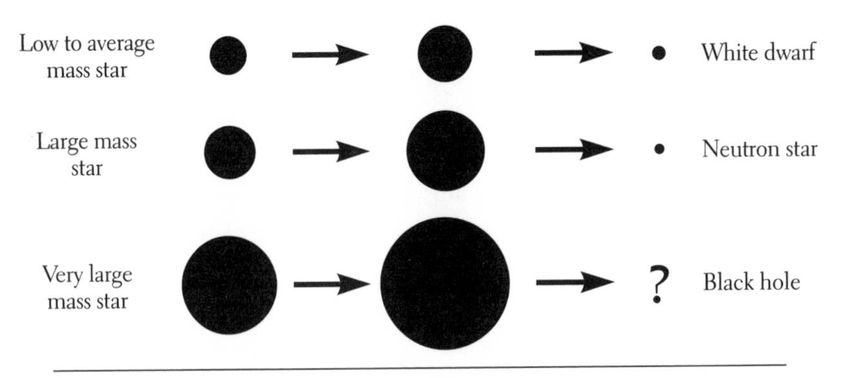

The endpoints of stellar evolution. (SAO/CXC/K. Kowal.)

sufficient matter to fall onto a neutron star or black hole and a shock wave to convert the energy of the infalling matter into heat, and you had an object that would produce far more X-rays than visible light: an X-ray star.

Where would the matter come from? A nearby star would do just fine. Roughly half of the stars in our galaxy are known to be part of double, or binary, star systems. If the stars in a few of these systems were sufficiently close together, then the gravitational pull of the collapsed star could steal some of the matter from its larger companion.

In 1962 Robert Kraft of the Lick Observatory in California proposed a model along these lines to explain nova outbursts of white dwarfs. Matter, consisting mostly of hydrogen since stars are mostly hydrogen, was pulled away from a close companion star to form a disk around a white dwarf. The matter gradually settled onto the white dwarf, building up a volatile store of compressed, hot hydrogen gas that eventually produced an outburst, by means of a thermonuclear explosion. The correctness of this model was confirmed by detailed calculations over a decade later, but the radiation from X-ray stars was steadier than could be accounted for by this mechanism.

Variations of Kraft's model were proposed to explain X-ray stars. These models called for a steady, vigorous flow of matter onto a collapsed star. The energy to explain the X-rays would not come from a nuclear outburst, which would presumably be suppressed by the heavy rain of matter onto the surface of the star, but through the release of gravitational energy. The model of Zeldovich and Novikov, and one by Iosef Shklovsky, another

Russian physicist, called for accretion onto neutron stars, whereas Kevin Prendergast and Geoffrey Burbidge, and independently Alistair Cameron and Michael Mock, preferred white dwarfs.

The white dwarf advocates questioned whether a binary system would survive the supernova explosion necessary to form a neutron star. The neutron star camp questioned whether accretion onto white dwarfs would produce a sufficiently strong X-ray source. Black holes were not seriously considered for two reasons: scientists were more confident of the existence of white dwarfs and neutron stars, and they were unsure whether matter would be heated as it fell toward a black hole or simply swallowed up without a trace.

In any case, accretion models required that X-ray stars be members of binary star systems, so the X-radiation should exhibit periodic variability as it orbited its companion. This was a clear prediction, but it was impossible to test with the technology of the day. The orbital periods of X-ray stars and their companions would take several days, but the only means of making X-ray observations was with rockets and high-altitude balloons, neither of which could stay aloft for that long. There were some conflicting observations of bright X-ray stars that hinted that some X-ray stars might be variable.

Observations of the X-ray stars with optical telescopes were also ambiguous. It was difficult to locate the positions of X-ray stars accurately enough to determine if any other stars might be associated with X-ray stars. Nevertheless, the binary star hypothesis remained a popular explanation, mainly because of its relative simplicity, until a startling discovery proved that neutron stars did indeed exist and that they emitted X-rays by a very different mechanism.

In late 1967, the British radio astronomers Jocelyn Bell and Anthony Hewish discovered a class of radio sources that blinked on and off with great rapidity and astonishing regularity. Pulsars, as these sources came to be known, were quickly determined to be neutron stars when one was discovered the next year in the Crab Nebula. The extreme regularity of the pulses—precise to a millionth of a second over a year or more—required that they be produced by either the rotation or pulsation of a stable object such as a star. The rapidity of the pulses, 30 times a second for the Crab pulsar, eliminated all types of star except a neutron star. Finally, the range of periodicities, from a few seconds to a few hundredths of a second, could be explained only by rotation. This theory was confirmed by observations of the very gradual slowing of the pulse rate, which showed that the pulsar was drawing its energy from rotation.

Herb Friedmans's group at the Naval Research Laboratory, with Richard Henry and Gilbert Fritz taking the lead, detected pulsed X-ray emission from the Crab Nebula pulsar in 1969. Shortly afterward an MIT group led by Hale Bradt, and a Rice University group led by Robert Haymes, confirmed and extended this discovery. Theorists immediately jumped on the pulsar bandwagon. The prototype for the model had been proposed in 1967 by Franco Pacini of Cornell University, who showed that a rotating neutron star with a large magnetic field would radiate large amounts of energy as very low frequency electromagnetic waves. These low-frequency waves would accelerate particles to high energies and produce intense beams of X-rays. Suddenly, gravitational energy was out and rotational energy was in as the explanation of X-ray stars. It seemed like a beautiful, simple explanation, but as Uhuru would soon show, it was wrong.

6

The Uhuru Years

THE UHURU YEARS, in the early 1970s, proved to be the adolescent wonder years of X-ray astronomy. The floods of exciting new observations and analysis were thrilling, but occasionally they led to sobering lessons in humility for those who scrambled to develop on-the-spot theoretical models.

Why did Uhuru make such a difference? In a word—time. In one day, Uhuru could scan the sky, rotating once every 12 minutes, or once every 60 seconds if necessary, for approximately 86,400 seconds. When we take into account times when the sun was too bright, or other efficiency factors, the amount of good observing time was about 50,000 seconds a day. A typical rocket flight yielded about 300 seconds of good data. So, in one day, Uhuru could accumulate the data equivalent of the information gathered on 167 rocket flights, and in one week it had accumulated more data than all the previous rocket flights in the history of X-ray astronomy. Further, Uhuru could make observations of a single source for a prolonged period time, which was impossible with a rocket.

A prime target for observation was Cygnus X-1. It was one of the brightest and most perplexing of the X-ray stars. Different groups of observers, including the group Harvey Tananbaum had done his Ph.D. thesis work with at MIT, had found ambiguous results for the intensity of X-rays from Cygnus X-1. Sometimes the results between groups agreed, and sometimes they did not. Either someone—more specifically, someone else, if you were one of the scientists involved—was making a mistake in sorting out the many effects that made the data messy, or the source was actually changing its intensity on some occasions and not on others. It was impossible to decide with the data at hand.

Uhuru observations of Cygnus X-1 twice in late December of 1970 and early January of 1971 settled the debate about the variability of Cygnus X-1. The intensity of its radiation was definitely changing on a time scale of weeks. Subsequent observations were scheduled.

Minoru Oda, who had taken a leave from the Institute of Space Science in Tokyo to visit AS&E for a few months to work on the data, quickly established that the intensity of Cygnus X-1 was varying rapidly, possibly periodically. On some scans through the source, the X-ray emission appeared to pulse regularly. On other scans it seemed that the pulse period had changed. Still other scans showed no evidence of regular pulsations. The data were confusing. One source of this confusion was the limitations imposed by the design of the satellite. Arrival times of incoming X-rays were recorded by accumulating counts for a fixed interval of time. The shortest accumulation time was 1/10th of a second.

The Uhuru data-recording system serves as a reminder of problems caused by designing even a few years ahead of time. Shortly after the system was approved, radio pulsars were discovered, and Herbert Friedman's group at the Naval Research Laboratory had observed pulsed X-rays from the Crab Nebula pulsar. The period of these pulses was 1/30th of a second. Since this was shorter than the accumulation time of the X-ray Explorer's data-recording system, Giacconi requested additional funds (about $250,000) from NASA to update the system to study the Crab pulsar and other similar objects that might exist. The request was denied, so the analysis of rapidly varying sources was severly hampered.

While Oda, Giacconi, Tananbaum, Paul Gorenstein, and Ethan Schreier—another recent MIT graduate who had come to work at AS&E—worked to understand the observational peculiarities of the data, another group struggled to fit Cygnus X-1 into the theoretical models of the day. This group included Bruno Rossi and Bruno Coppi at MIT, Oda, and me (WT). Oda had developed a working hypothesis that the source was pulsing with a period that was less than the data accumulation time of 1/10th of a second. If this was true, then it would be difficult if not impossible to determine the period uniquely, but a pulse period of 73 milliseconds was consistent with the available data. This was reasonably close to the Crab Nebula pulsar period of 33 milliseconds. Such rapid rotation posed problems for accretion models, so we began working on models based on the rotating neutron star model that explained the Crab pulsar so well.

According to this model, a neutron star should have an intense magnetic field that, like the rapid rotation, was a consequence of the collapse of the stellar core that became a neutron star. Since the time of Michael Faraday, it had been known that a rotating magnetic field produces electric fields—this is the basis of electric generators. The rapid rotation and intense magnetic field of a neutron star would produce stupendous electric fields, and a blizzard of high-energy particles. As these particles stream away from the neutron star, they radiate to produce radio, optical, and X-ray pulses. The striking similarity of the X-ray spectrum—the distribution of X-rays with energy—of Cygnus X-1 to that of the Crab pulsar gave us further confidence that we were on the right track.

I was enthusiastic about the rotation-powered neutron star model for a number of reasons. My thesis work had been on models for X-ray emission from the Crab Nebula, I had been an officemate of Franco Pacini at Cornell when he had his ingenious insight that led to the rotating neutron star model for generating energy for the Crab Nebula, and I had worked on similar models for X-ray stars at Rice University and later at AS&E. As it turned out, this background did not serve me well in the months that followed.

There was a major problem with the Crab pulsar analogy. If Cygnus X-1 was like the Crab pulsar, why was there no evidence for an extended cloud or nebula around Cygnus X-1 such as existed around the Crab pulsar? One feasible answer was that Cygnus X-1 was much older than the Crab pulsar, so the nebula had faded away. This implied that a neutron star in Cygnus X-1 would have had a larger store of rotational energy than the Crab pulsar. This line of reasoning led to a model in which Cygnus X-1 was a massive neutron star supported by its rapid rotation.

Giacconi, who would pop in on our discussions at Rossi's office from time to time, was disappointed with this model. He wanted Cygnus X-1 to be a black hole. We patiently explained to him that a black hole could not produce regular pulsations, because it has no rigid surface. Any pulses produced by matter before it falls into a black hole would be erratic and short-lived.

When I presented the massive, rapidly rotating neutron star model for

Cygnus X-1 at a conference, scientists and reporters alike were also disappointed. They, too, wanted Cygnus X-1 to be a black hole. "Sorry," I said. "I don't see how that is possible, given the regular pulsation period."

A few months later a paper describing the observations was prepared and submitted by Oda and his colleagues. In it, they cautiously stated that "we cannot exclude the possibility that the period is a multiple of 73 milliseconds," and added, at Giacconi's insistence, a comment that Cygnus X-1 might be a black hole.

While X-ray groups at MIT and the Goddard Space Flight Center planned rocket flights with fast data recorders to search for the 73-millisecond period, Giacconi, Tananbaum, Schreier, and others were staying up late at night and working on weekends, sifting through the data on other X-ray sources for evidence of periodic variations. They found one within a matter of weeks. Centaurus X-3, an X-ray star in the constellation Centaurus, was emitting pulses of X-rays every 4.8 seconds. The pulsations were much more regular than those from Cygnus X-1, but not so regular as those from the Crab pulsar. Cygnus X-1 was put on the back burner, and the observing schedule was hastily revised to accommodate further observations of Centaurus X-3.

For my part, the observations of Centaurus X-3 were a clear signal that my rotation-powered neutron star model for X-ray stars should be put not on the back burner, but rather on the compost heap. The rotation period of 4.8 seconds for Centaurus X-3 was too slow to provide any effective generator action for more than a few years, an unacceptably short time unless one was willing to believe that we were witnessing the formation of this object. This was a very remote possibility in view of the existence of dozens of other X-ray stars. It was much more probable that Centaurus X-3 was either a pulsating white dwarf or a rotating neutron star that was accreting matter from a companion.

In May of 1971, brilliant observational work by Schreier, Giacconi, Tananbaum, and Richard Levinson established that Centaurus X-3 was orbiting a companion star every 2 days. Wojciech Krzeminski of the Carnegie Institute of Washington located the companion star with an optical telescope. It was a massive blue supergiant star with a mass of about 20 suns, and showed evidence for a 2-day orbit. In July, another X-ray star, Hercules X-1, was found to have a pulse period of 1.24 seconds and an or-

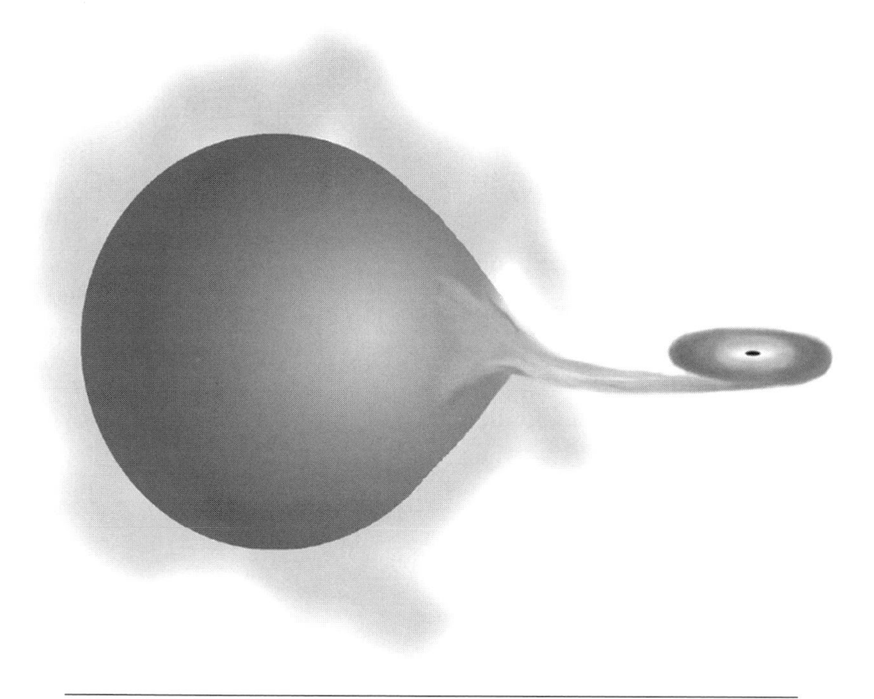

Schematic of X-ray emission from a neutron star or black hole accreting matter from a companion star. (SAO/CXC/S. Lee.)

bital period of just under 2 days. This discovery effectively eliminated the pulsating white dwarf models.

What of Cygnus X-1? Was it also an accreting neutron star, or could the rotation-powered neutron star model still be salvaged for it? After all, as Giacconi had pointed out, Cygnus X-1 was different. The observations of the MIT and Goddard group, and a reanalysis of old data by the Naval Research Laboratory group, proved fatal for the neutron star generator model. The proposed 73-millisecond period did not exist. The X-ray intensity fluctuated very rapidly, as often as 20 times a second, and the fluctuations were random and erratic rather than periodic. It was beginning to look as if Giacconi's hunch might be correct.

A concerted effort by the Uhuru team, and by other X-ray astronomers, especially the MIT group, and by optical and radio astronomers, led to the conclusion that Cygnus X-1 is part of a binary system. Detailed studies of Cygnus X-1 and its companion star by Charles Bolton at the University of

Toronto and others showed that Cygnus X-1 must be a collapsed star with a mass of about 10 solar masses. That, according to the current understanding of nuclear physics, stellar structure, and Einstein's theory of gravity, would make Cygnus X-1 a black hole.

The mystery of X-ray stars had been solved with stunning swiftness, largely because of the increase in the monitoring capability of Uhuru. They were slowly rotating neutron stars or black holes in which X-rays were produced by the gravitational accretion of matter from a nearby companion star. It was, as Giacconi put it simply and eloquently, "the most mystical moment, when we suddenly understood."

Uhuru was a watershed in the history of astronomy. The earlier discovery of pulsars and the Crab Nebula pulsar in particular by radio astronomers had established the existence of neutron stars. But except for the X-rays emitted by the Crab, which were clearly a short-lived special case, X-rays generated by pulsars were insignificant. The identification of X-ray stars as accreting neutron stars and black holes demonstrated three important aspects of our universe.

First, it established that our galaxy contains thousands of collapsed stars that radiate most of their energy in X-rays, and blink on and off in a fraction of a second.

Second, in the case of Cygnus X-1, it provided the first substantial observational evidence for the existence of black holes. This discovery is the one that has captured the popular imagination. The concept of black holes has now become part of the popular literature as a metaphor for an irretrievable loss of material, time, money, information, and so on.

Finally, it showed that gravitational accretion is an important process and source of power in the universe. Although this discovery does not have the glamour attached to black holes, it has broader implications for practicing astrophysicists. Gravitational accretion is now invoked to explain a wide range of phenomena besides X-ray stars, from the formation of disks around very young stars, to white dwarf stars that show explosive activity, to the enigmatic and powerful quasars.

After the observations, and the theoretical work of the Russian astrophysicist Rashid Sunyaev and the English astrophysicist Martin Rees and their collaborators on accretion disks in X-ray binaries, it became possible to accept that the same process could explain the awesome energy produced in the center of some galaxies and quasars. Ironically, even as X-ray binaries have become only one of many examples of the accretion process, their study has gained in significance. Because of their brightness and rapid time scale of variability, they are excellent cosmic "laboratories" in which to test theories of the accretion process.

The accomplishments of Uhuru were not confined to the densest states of matter in the universe. While Giacconi, Tananbaum, Schreier, and others were concentrating on X-ray stars, Herbert Gursky, Ed Kellogg, and Steve Murray, who joined AS&E after receiving his Ph.D. from the California Institute of Technology, showed that distant clusters of galaxies were also intense sources of X-rays. It would take the focusing power of the Einstein Observatory to more fully explore the major scientific implications of this important discovery.

7

The Einstein Observatory

UHURU'S SCIENTIFIC successes could not have been better timed politically. The original proposal for a large X-ray telescope had evolved into a proposal by a consortium of institutions—AS&E, Columbia University, the Goddard Space Flight Center, and MIT. The centerpiece of their proposal, the Large Orbiting X-ray Telescope, was an X-ray telescope with five sets of mirrors, each approximately 36 inches in diameter. Their proposal was accepted for one of the four missions scheduled for the High-Energy Astronomy Observatory (HEAO) program in 1970, but major hurdles remained.

They were third in line, and only the first two missions were fully funded. An X-ray telescope was still viewed as a risky venture into unproven technology that needed more study. Two more years passed without approval. Meanwhile, the winds of an economic recession were blowing the HEAO program further off course.

The Nixon administration put a ceiling on spending by executive agencies. NASA had to cut its budget at a time when it was staggering under the burden of cost overruns for the Mars Viking program. Extra money, a lot of extra money, had to be found within the NASA budget to keep the Viking program on target for a Mars landing in the bicentennial year of 1976. The only program with that kind of money was HEAO. On January 2, 1973, the HEAO program was canceled without warning.

Fortunately, the HEAO program had an able advocate in Jesse Mitchell, head of NASA's physics and astronomy division, and a savvy veteran of NASA politics. He knew that the NASA bureaucracy was like a big ocean liner. A small change in direction now could make a large change ultimately. He sought a change that had negligible financial implications for the moment, but had enormous scientific consequences for the long run: change the status of the HEAO program from "canceled" to "suspended." The discoveries made by Uhuru were crucial to his case. The astronomical community rallied to the cause. NASA Administrator James Fletcher agreed, and gave the NASA program managers and scientists 18 months to come up with a detailed workable alternative program costing half as much, or face cancellation.

Three missions emerged from the restructuring. HEAO-1, in essence a much larger version of Uhuru, would survey the sky over a wide range of X-ray energies. HEAO-2 would be a reduced version of the large X-ray telescope proposed by Giacconi's consortium. It would have four sets of mirrors rather than five, and the mirrors would be half the size of those originally proposed. The third HEAO mission would carry instruments designed to detect gamma rays and cosmic rays.

A smaller, revised version of the large X-ray telescope was not what Giacconi and his colleagues had hoped and planned for, but they were consoled by two major improvements in their situation. This telescope was funded, and it was second in line. The launch date for HEAO-2, which would later be renamed the Einstein Observatory, was set for late 1978.

During the restructuring of the HEAO program Giacconi decided to leave AS&E. Prompted by changes in the focus and policies of the company, he moved across Cambridge to the Harvard University campus, where he became associate director of the high-energy astrophysics division of the Harvard-Smithsonian Center for Astrophysics. Many members of the high-energy group moved with him, and so the excellent team he had assembled for the Uhuru program remained largely intact.

Giacconi, as the principal investigator for the Einstein scientific consortium, had overall scientific responsibility for the entire observatory. Herb Gursky, Giacconi's long-time colleague, moved to the Center for Astrophysics with the group, but switched fields to become associate director of the optical and infrared division. Harvey Tananbaum, who had impressed Giacconi with his ability to get things done on the Uhuru program, became scientific program manager for the Center for Astrophysics part of the consortium. This part included responsibility for the development of two X-ray detectors and the X-ray mirrors.

The four pairs of X-ray mirrors were the most crucial observatory components. No redundancy was possible; not even a test mirror was allowed. The mirrors had to be ground, polished, and coated to a tolerance of one ten millionth of an inch, on schedule and for the lowest possible cost. The primary responsibility for this awesome task was entrusted to Leon van Speybroeck. He had worked with Giacconi, Giuseppe Vaiana, and others to develop an X-ray mirror assembly that flew aboard Skylab and produced spectacular X-ray images of the sun.

Van Speybroeck is a master mirror-maker in the tradition of Herschel, with a modern twist. Today's mirror-makers do not do all the grinding or polishing, nor do they have their sister do it, as Herschel did. Van Speybroeck uses computer ray-tracing programs and laboratory experiments to test the design, while the actual grinding and polishing are subcontracted out to a firm with highly skilled scientists, engineers, and technicians. Perkin-Elmer Corporation in Danbury, Connecticut, was the subcontractor for the HEAO-2 mirrors, but the mirrors were always on van Speybroeck's mind, and often in his meticulous, demanding sight.

"Leon hovered over those mirrors like a mother hen," Giacconi recalled.

"He literally slept with them," Harvey Tananbaum said. "He never let them out of his sight during the crucial periods."

Van Speybroeck was in constant contact with Peter Young, who was the scientist in charge of the project at Perkin-Elmer. With Young's cooperation, van Speybroeck pushed for ever more precise polishing of the mirrors and better testing procedures at every step of the process, pressing the limits of existing technology and trying NASA's patience. At more than one stage NASA intervened at what van Speybroeck and the mirror team considered arbitrary schedule milestones to force them to move on beyond what NASA considered excessive striving for perfection. Van Speybroeck's reaction: "If I were doing it over, I would want to control certain environments even better."

Van Speybroeck's oversight prevented countless small mistakes and several large ones. One of the more dramatic examples occurred during the loading of the mirrors. The crate containing the mirror assembly was to be loaded by a crane onto a truck. The ever cautious van Speybroeck insisted that the loading process be tested in a dry run, with a crate containing ballast. On the first run, the crane dropped the crate! After much checking and rechecking of the loading procedure, the mirror assembly was safely loaded.

No matter how good the mirrors were, they would have been of little use without a detector that could faithfully and precisely record the position of the focused X-rays. Paul Gorenstein was building an instrument based on the tried and true proportional counter technology. It would turn out to be an extremely useful detector, but it could provide only broad-brush images.

An entirely new type of detector, one that could fully exploit the capabilities of the mirrors, also needed to be developed. The proposed detector was based on a vidicon system, basically a 1970s version of a TV camera, but it proved too difficult to adapt to an observatory in orbit. A group of scientists that included Steve Murray and Pat Henry then set off in another direction, and began studying a modification of a detector developed by scientists at the University of Leicester in England. This led to a design for a detector they believed they could build in time and on budget, the High-Resolution Imager.

The High-Resolution Imager, which became the model for a similar instrument that would be used years later on AXAF, was about the size of a 35-millimeter slide. It consisted of two sets of about 10 million tiny glass tubes, each with a diameter about a fourth that of a human hair. These tubes were especially coated with a material that converts X-rays into electrons. When an incoming X-ray hits the side of one of these tubes, it initiates a cascade of electrons. After the electron cloud emerges from the second set of tubes, it hits a crossed grid of wires and produces an electric signal. This signal can be analyzed to reconstruct an image of the cosmic source of the X-rays.

Murray and his colleagues sought help from high-tech companies to build the High-Resolution Imager, but finally decided to build it in the basement of the Center for Astrophysics.

"We just kept trying things until it worked," said Murray. "After going down many blind alleys, finally we succeeded. In the end, it turned out that the simplest ways were the ways that worked." For example, how do you get 128 wires to the inch, evenly spaced? Wind a double strand of wire and then unwind one strand.

It seemed as if everywhere they turned, a new set of problems arose, especially with the mirror assembly. "It was a bigger and more difficult mission than everyone had estimated," said Albert Opp, who was the NASA program scientist for the HEAO program. "The optical system was an exceedingly sophisticated, exceedingly complicated system."

Tananbaum agreed. "We had to solve a hundred thousand technical problems."

Did they doubt they could do it?

"There were periods of nervousness," Giacconi allowed, "but never panic."

Evidently not. In June of 1975, they were still struggling to devise a workable detector and mirror designs. Even so, Giacconi wrote a letter to NASA stating his intent to submit a proposal for the large X-ray telescope denied them by the HEAO budget crisis.

Six months later, he and Tananbaum submitted the proposal in the form of a long letter to Noel Hinners, who had replaced John Naugle as the Associate Administrator for Space Science at NASA. The large telescope would have six nested sets of mirrors, with the biggest mirror being 46 inches in diameter. Hinners responded that he was interested. Giacconi and Tananbaum followed up with a more detailed proposal.

In August 1977, funds were approved for a feasibility study in coordination with the Marshall Space Flight Center for "the Advanced X-ray Astrophysics Facility," or AXAF, as the telescope was called. Why call it a facility instead of a telescope? Because the proposed Large Space Telescope, which would become the Hubble Space Telescope, was proving a hard sell both within NASA and on Capitol Hill. Congress might not be ready to fund two large telescopes in space, the reasoning went, but it might fund one telescope and one "facility." Incredibly, the strategy worked, at least to the point of getting funds for a 3-year feasibility study.

Twenty-two minutes after midnight on November 13, 1978, the Einstein Observatory was launched into orbit with an Atlas-Centaur rocket. Four days later, the High-Resolution Imager was moved into the focus and the telescope was pointed toward Cygnus X-1. Giacconi watched on a monitor at the Goddard Space Flight Center as the telescope scanned toward Cygnus X-1. X-ray images of stars streaked across the field of view. The telescope fixed on Cygnus X-1 and a point-like image formed on the screen. Eighteen years after the publication of the article with Rossi on the feasibility of an X-ray telescope, Giacconi knew he had a working X-ray telescope in orbit. "It was an unbelievable sensation," he said, "a profoundly moving moment."

For van Speybroeck, the first X-ray image was "almost like a religious experience."

The Einstein Observatory brought the X-ray sky into focus. Almost immediately, spectacular X-ray images of supernova remnants and clusters of galaxies showed the value of an X-ray telescope over simple detectors. The X-radiation from the outer layers, or coronas, of stars such as the sun was detected for the first time, at a level 100,000 times weaker than X-rays from SCO X-1, the first source discovered. Observations of supernova remnants using a detector built by the Goddard Space Flight Center provided clear evidence that heavy elements are ejected into space by supernovas. The High-Resolution Imager captured the tableau of the awesome shock waves that spread these vital-to-life elements—carbon, nitrogen, oxygen, and others—through the galaxy.

The Einstein Observatory also established that the most powerful X-ray sources are quasars, or quasi-stellar objects. Quasars are so far away that some astrophysicists doubted (a few still doubt) if the method used to determine their distance (the red shift) was correct. If quasars are several billion light years away, as the measurements imply, they are radiating energy at a rate equal to a thousand or more galaxies from a compact region less than a light year across. The X-ray observations by Einstein lent strong support to the theory that the awesome power of a quasar is produced by clouds of gas falling into a supermassive black hole in the center of a galaxy. Estimates indicate that the central black hole can contain the mass of several billion suns.

At the other extreme, X-ray images of hot gas on the outer edges of galaxies and in the space between galaxies confirmed a growing suspicion among astronomers that something important is missing in their inventory of galaxies and clusters of galaxies.

In the 1970s astronomers used improvements in optical and radio light detectors to gather data on subtle shifts in the frequency of light waves from stars and gas clouds on the faint outer edges of spiral galaxies. By interpreting these shifts as due to the motions of the stars and clouds producing the radiation, they obtained measurements of the speeds of rotation of the galaxies. Astronomers could estimate what the rotation speed of a galaxy should be by calculating the mass of all the stars and gas in the galaxy, which determined the gravitational force of the galaxy. They were surprised to find that most galaxies are rotating much faster than they should. Stars and gas clouds on the outer edges of spiral galaxies should be flying away from the galaxies like droplets of water from a rotating sprinkler head. Yet there they are, rotating majestically as if nothing is wrong.

Years ago, Fritz Zwicky of supernova and neutron star fame, had pointed out that a similar problem existed for clusters of galaxies. Although the motion of galaxies in a cluster appears more like a swarm of bees than a graceful spiral, it still has the look of motion in equilibrium with the gravitational force of the cluster. The cluster is not flying apart. Yet, as with individual galaxies, an inventory of the combined mass of all the galaxies in the cluster did not account for enough gravity to hold the cluster together.

Einstein X-ray observations brought important independent evidence to bear on this problem. The measured pressure force of the hot gas in clusters of galaxies was greater than the measured gravitational force, so the gas should have expanded out of the cluster like air from a leaky balloon. Yet Einstein showed that the hot gas was there, in cluster after cluster.

The most popular solution to this dilemma is to postulate that most galaxies and clusters of galaxies are enveloped in some form of "dark matter" that has yet to be observed by radio, infrared, optical, ultraviolet, X-ray, or gamma ray telescopes. The consensus is that between 3 and 10 times as much mass is in dark matter as in all the stars and gas. This means that most of the matter in the universe is still unseen. Such a shortfall would get a clerk at a bank or a warehouse arrested. Fortunately, astronomers are not held to the same standards, and rightly so, since they were not around when the universe was put together.

With Einstein, X-ray astronomy became part of the mainstream of astronomy. By reserving a significant amount of observing time for guest observers and by eventually making the data public, the Einstein program attracted specialists on other wavelengths to X-ray astronomy. This cross-fertilization reinvigorated astronomy as a whole. What was needed now was a large X-ray telescope that could send back images of a quality equal to that of the best optical telescopes.

II

A NEW CONCEPT AND A NEW START

8

Ready for the Job

"I KNEW THAT THIS was the job that my life had been getting me ready for doing."

The speaker was Charlie Pellerin, and he was talking about his appointment in February of 1983 to be director of the astrophysics division at NASA Headquarters. He served in that capacity from February 1983 to June 1993. His tenure was twice that of any of his predecessors or successors. During this time his position as astrophysics director and his knowledge of the system gave him a strong hand to play in determining AXAF's fate.

We were sitting in our sock feet in the living room of his "no-shoes-allowed" home—a reflection of his interest in Japanese culture, which is confirmed by the paintings on the walls—in Boulder, Colorado, where he settled after leaving NASA.

Pellerin's route to the job he had spent his life preparing to do was not a direct one. But neither was it a random walk. His father, a real estate sales agent, had sold a house to a man from NASA. Like any proud father, he mentioned to his client that his son had just received a bachelor's degree in physics from Drexel University in Philadelphia. The man suggested that his son apply for a job at NASA's Goddard Space Flight Center in nearby Greenbelt, Maryland. Shortly after that, Charlie Pellerin applied for and got a part-time job there that allowed him to finance his way through graduate school at Catholic University.

After completing his doctorate in physics Pellerin went to work in the laboratory for high-energy astrophysics at the Goddard Space Flight Center on a project in cosmic ray physics. He soon realized he was in the wrong place.

"You know what it's like," he said. "You do integrations over this and get the units right and the four pi's right. You're in offices where the phone rang three times a day. Once was my wife and twice was the wrong number. We just worked all day on these calculations."

It was at this time that he became aware of X-ray astronomy, which was capturing many headlines in astrophysics because of Uhuru.

"It was popular to hate X-ray astronomers," he said. "Because the X-ray astronomers seemed to have all the visibility, all the money, and we were working so hard . . . It was the most onerous task I've ever done. I hated it. I was never any good at, like detail work. I hated doing calculus problems as an undergraduate and here I was again, but these were real ones. We quit at about six at night, typically, and I would go home to my wife and two kids and watch TV or something and come back the next morning, and I thought, oh my god, we got to do this work again today."

One morning he came in to work and met a colleague, who said, "Guess what I did last night. I had the time of my life."

"And I say," Pellerin recalled, "No kidding, what did you do? He says, 'Well, I went home, drove in the driveway, met the wife and kids, had a real quick dinner, drove into my office, locked the door and finished these calculations. Worked to three in the morning and finished these calculations. I feel great.'"

"I said to myself, these are not my people. I need a different kind of work. This is not fun for me. That's really the day I decided to go do something else."

Pellerin bounced around from one job to another within NASA for the next 4 years, gradually moving away from science into management. First he worked in planning payloads for the Space Shuttle until that organization was "annihilated in a typical NASA battle." He wound up in another division in the Office of Applications managing the development of payloads for the Shuttle. Then he joined forces with a colleague to form yet another division within the Office of Applications. What looked like a good idea turned bad when a tragic mountaineering accident took the life of Tim Mutch, the Associate Administrator of Space Science and Applications and precipitated yet another reorganization.

"I kinda ended up in a limbo space and Frank Martin rescued me from a really bad situation." Martin, who was director of the astrophysics divi-

Charles Pellerin. (NASA.)

sion of the Office of Space Science and Applications, made Pellerin his deputy director.

"So, I didn't come into the job through any kind of a career plan," Pellerin said. "Someone just put me there."

Where he had been put was in the middle of a division under siege. Technical and management problems had put the Hubble Space Telescope way over budget and far behind schedule. An infrared astronomy satellite was also costing several hundred percent more to build than budgeted.

"It was a really difficult time," Pellerin said. "Frank Martin was getting a constant stream of harassment aimed at humiliating him about the Hubble problems . . . some [NASA officials] were acting out their frustrations by terrorizing us and we were probably doing everything we could to aggravate them. To say it wasn't much fun would be the understatement of the year."

Pellerin felt that he had not been adequately prepared for this type of intra-agency guerrilla warfare.

"I realized that I had trained something like 11 years to be a research physicist and now was sort of a NASA executive and my training was about 2 days. There seemed to be an imbalance."

He took a 4-month leave of absence to attend Harvard Business School's program for management development, a program for people at the upper level of middle business management, to teach them "how to manage MBA's."

"In a way it was kind of an odd program because it had to do with a lot of things that weren't government policy," Pellerin explained, "but almost 90 percent of NASA's budget is spent in industry."

Pellerin immersed himself totally in the program.

"We worked day and night, with only Saturday nights off. After 2 weeks I forgot all about NASA and I didn't contact anybody."

He reported back for work at NASA the day after New Year's in 1983, now confident that he could be a strong and resourceful ally to Martin.

"So I walked into Frank's office and started trying to learn what had happened while I was gone. And he said, 'IRAS [the Infrared Astronomical Satellite] is going to be on the launch pad in another 30 days, Space Telescope has major budget problems, I'm leaving, and I'm recommending that they put you in charge.'"

A stunned Pellerin asked Martin what his timetable was.

"I have bled for IRAS," Martin said, "and I'm going to see it fly. It's going to launch in 30 days; then I'm out of here. In the meantime you can do whatever you want around here."

One of the first things Pellerin learned was that the budgetary and scheduling problems with the Hubble Space Telescope had not gone away. The projected overrun had risen to $400 million and the schedule was slipping on an almost monthly basis.

"It was a big embarrassment to the agency," Pellerin said. "So I spent the next 4 months almost completely involved with Hubble problems, and a lot of it flying around the country with Jim Beggs [then the head of NASA], visiting the contractors."

NASA set up a "tiger team" of specially recruited experts to review the management of the Hubble Space Telescope. One result of their review was the reassignment of Fred Speer, the project manager at the Marshall Space Flight Center. In Pellerin's view, Speer, who had been the project manager for the Einstein Observatory, was a scapegoat.

"What was really the matter with Hubble," Pellerin said, "was not Fred. It was a systemic problem in the culture that permeated the whole thing."

The Hubble trouble affected more than just Speer.

"A whole staff of people got slandered in this," Pellerin related. "And the tragedy of that was that they were trying to tell me about these problems that summer [1982] while I was there and I really wasn't fully hearing what they said. There were so many stories going around and it was so complex."

Pellerin figured that the only reason he survived was that he had been out of town during the worst part of the Hubble management problems. And with the worst part over, Pellerin was ready to go to work.

"It was one of those rare moments in life when I've walked into something that was really so big, as director of astrophysics . . . It's the first job I've ever had that I walked into and knew—this is my job. I knew deep down that somehow I was going to have that job, and I could do it right. I just knew that. With the management training I had done, my scientific background, my talents just seemed right."

If he felt right for the job, Pellerin also felt that it was an important job that needed to be done. In spite of all its problems and the at-times vicious infighting, he liked working for NASA.

"I believed deeply in NASA's mission, and in the importance of physics and astronomy, and I took this charge as an incredibly serious thing to make the most of this. I always thought that this era is a special one and not likely to be repeated anytime soon."

9

Blazing the Trail

Any mission the size of AXAF must travel a long and treacherous road before it arrives at a NASA launching pad. Before the satellite carrying the dreams of its creators lifts off the pad, the dream must have been shared by tens, then hundreds, then thousands of people—scientists, NASA officials, industry management, engineers, technicians, congressional staffers, members of Congress and the executive branch, and ultimately the voters, who may be called on to express their support of NASA in general and a specific mission in particular.

The first part of the trail must be blazed by the individual or group that conceives the mission. They—it must very quickly become "they," not just one person, who are proposing the mission or it will be stillborn—formulate a concept, such as a large X-ray telescope, and begin to sell it to their scientific peers. This is done by making personal appearances—preferably as invited speakers, a cachet that lends important credibility—at meetings of the American Astronomical Society or the International Astronomical Union, and at meetings of the scientific advisory committees of NASA and the National Academy of Sciences.

Riccardo Giacconi was on the road early and often in his advocacy of AXAF. In 1963, pressing the advantage of his group's discovery of the first X-ray star, he submitted a proposal for a large X-ray telescope to NASA. As a member of the X-ray and gamma ray panel of the Space Science Board on the National Academy of Sciences that met in Woods Hole, Massachusetts, in 1965, he pushed for and won approval of the concept. Two years later, he secured the recommendation of NASA's astronomy missions board, and in 1970 he had put together the powerful consortium of American Science & Engineering, Columbia University, the Goddard Space

Flight Center, and MIT that eventually built the Einstein X-ray Observatory. While Einstein was under construction, he and Harvey Tananbaum began working the same ground to build support for AXAF.

Once a mission has been approved for study by NASA, it is out of the woods and onto a well-traveled road that is littered with abandoned or wrecked dreams of good and cherished missions that might have been. AXAF reached this road in 1977 with NASA's approval of funds for a conceptual design and preliminary analysis of the mission.

Conceptual design and preliminary analysis are called Phase A studies. They represent the first of five well-defined phases for any NASA mission. These are Phase A; Phase B, the detailed definition studies; Phase C, design; Phase D, development; and, finally, the operations phase. The timetable for progressing through these phases is highly uncertain, and approval of funds for one phase is no guarantee that the mission will go on to the next phase, or even that it will make it through that phase without cancellation.

Phase A studies for AXAF began in June 1977, in response to the 1976 proposal from Giacconi and Tananbaum. The proposal also included the concept that a national X-ray astronomy institute be established to manage the mission, along with the Einstein Observatory.

"The idea of an institute made NASA nervous," Tananbaum recalled. "The concept of our group doing the whole mission wouldn't sell."

The Harvard-Smithsonian group had developed a working relationship with the Marshall Space Flight Center while building the Einstein Observatory, so they joined with them for a successful AXAF proposal. The Marshall Space Flight Center would be responsible for the overall management and systems engineering, and the Harvard-Smithsonian group would provide scientific and technical support to the project scientist at Marshall. Although the Harvard-Smithsonian group was initially disappointed that they could not run the whole AXAF program, they soon came to see the value in the partnership.

"It was a healthy development," Tananbaum acknowledged. "The teaming of our science and engineering capabilities with Marshall's skills in managing large space programs has been a powerful and positive aspect of the program."

The cooperation between the teams at Harvard-Smithsonian, Marshall,

MIT, Penn State, and the many contractors in industry distinguished the AXAF mission from that of the Hubble Space Telescope.

"Hubble didn't have this—a mission support team helping to keep the scientific interests in mind," Tananbaum explained. "No group on Hubble had end-to-end support on the project with the kind of staff we had. We had a staff of a dozen experienced scientists and a dozen experienced engineers who were outspoken because they had previous experience on a mission like this. Marshall and the contractors didn't love us all the time but they respected us because we didn't hesitate to speak our minds. That they didn't have on Hubble."

Indeed, the report of the Allen Board of Inquiry that looked into the Hubble mirror debacle emphasized that, for Hubble, "the NASA Scientific Advisory Group did not have the depth of experience and skill to critically monitor the fabrication and test results of a large aspheric mirror." The board's report went on to criticize NASA's inability "to penetrate the process and ask for validation."

"The scientists on AXAF had real hardware experience and a culture of end-to-end testing as the thing to do," Tananbaum said. "We wouldn't have accepted 'No' for an answer from NASA. It was the thing to do and we were going to do it. We would either have found a less expensive way to do it, or we would have forced it to the very top of NASA and done it."

Tananbaum smiled at this point, realizing that he might be exaggerating the power of the scientists. "We don't always win," he corrected himself, "but it would have been a hell of an interesting wrestling match to see how it played out."

The concept of a strong mission support team was not a lesson learned from Hubble. Hubble had yet to experience its problems by the time the Harvard-Smithsonian and Marshall partnership was formed. Instead, this concept was ingrained in X-ray astronomers from the beginning as the way to approach missions because, from the beginning, they had to do their projects in the unforgiving environment of space.

Fred Wojtalik, the AXAF program manager at Marshall, who had served in the same role for Hubble, was well aware of this difference in approach.

"This culture [the AXAF team] is easier to work with," he said. "I'm not sure what it is, but they have helped us tremendously, when we make

Martin Weisskopf. (NASA/MSFC.)

trades and try to make ends meet . . . They are much more members of a team."

Part of this compatibility was due to relationships formed during the years many of the key people worked on Einstein together.

"Giacconi, Tananbaum, van Speybroeck, Murray—we were old friends."

Wojtalik also credited Marshall's AXAF project scientist, Martin Weisskopf, for the cooperation between the Marshall staff and the scientists working on AXAF.

"Martin has been tremendous. He is very knowledgeable, very respected by the scientific community. I feel I am very lucky to have a person of his caliber in that position."

Once it was established that Marshall would manage the AXAF program, it became clear that Marshall should hire a scientist recognized in the field of X-ray astronomy who could speak and act on science matters from the Marshall perspective and serve as a direct channel of communication for other scientists to the project manager.

They chose Weisskopf, who had graduated 8 years earlier from Brandeis University with a Ph.D. in atomic physics. After Brandeis, he went to Columbia University to do postdoctoral research with Robert Novick, who had set up an X-ray astronomy group. In a sense, Weisskopf began working on AXAF at that point. One of his first jobs was to help with Columbia's portion of the 1970 AS&E-Columbia-Goddard-MIT consortium proposal for a 1.2-meter X-ray telescope. That proposed X-ray telescope was the one that was downsized a few years later and flew in 1978 as the Einstein Observatory. A year earlier, Weisskopf had left Columbia for Marshall to become project scientist for AXAF, a position he has held for more than 20 years.

The challenges in the early years, Weisskopf remembered, apart from dealing with the strong personalities associated with the project, were mostly political. AXAF simply was not a high priority for upper-level NASA management at the time. "Getting recognition and funding for the program was the principal challenge . . . We were all convinced that the mission was the scientifically logical follow-on to Einstein—very obvious after the Einstein results came pouring in."

The political problems were not with the perceived scientific merit of AXAF. They were with the Hubble Space Telescope.

"The effects of Hubble's cost, growth, and management problems were devastating," Weisskopf said. "AXAF, being an astronomy payload, was considered to be, one, behind Hubble, and two, identical to Hubble."

The latter comparison led to misleading and potentially deadly cost estimates for AXAF. Every time the Hubble cost went up, AXAF's projected costs were questioned. Weisskopf found himself trying to persuade NASA to reduce the cost estimates, which he felt made the mission look too expensive to accomplish. Part of the problem was the difference between the Hubble and AXAF scientific teams.

"The HST [Hubble Space Telescope] science community was very different from ours," Weisskopf explained. "Project science for AXAF consisted of many people, not only myself, and the scientists at SAO [the Smithsoniam Astrophysical Observatory] that were part of the project, but also the group that I had been building [at Marshall]. HST on the other hand was run with one project scientist and a deputy—a reflection of a different way of doing business and a lot of naivete."

This showed up, as Tananbaum remarked and Weisskopf agreed, in the unyielding attitude of the Hubble scientists on the requirements and their lack of involvement with the engineers.

"The engineers, by and large," Weisskopf observed, "prefer that the scientists state requirements and then go away. Those more experienced realize that this approach cannot work. X-ray astronomers simply learned earlier, in the sixties."

An AXAF science working group was set up by NASA to assist in establishing scientific requirements and to give the benefit of their experience to the Marshall and Harvard-Smithsonian teams during the conceptual definition studies. Giacconi was the chairman of the group, Weisskopf was vice chairman, and Tananbaum was a member, along with 13 other scientists from NASA, universities, and research institutions around the world. In their report published in 1980, they described the preliminary design of AXAF. It would consist of six nested pairs of mirrors. The largest pair of mirrors would be 1.2 meters in diameter, and the telescope would have 60 times the focusing ability of Einstein and 100 times the sensitivity to faint sources of X-rays. It would be launched on the newly developed Space Shuttle. The projected launch date was mid-1987, assuming that they got the go-ahead from NASA for definition studies (Phase B) immediately and for design and development (Phases C and D) by 1983.

In their report, the AXAF science working group described the considerable scientific promise of AXAF, and predicted that it would be "an extremely powerful and unique tool for new and fundamental discoveries in astrophysics." Their colleagues in astronomy agreed. In 1982, the influential Astronomy Survey Committee of the National Research Council,

Riccardo Giacconi (left) and Wallace Tucker (right) at the Space Telescope Science Institute in 1983. (K. Tucker.)

chaired by George Field, director of the Harvard-Smithsonian Observatory, released its report *Astronomy and Astrophysics for the 1980's*. The committee of two dozen experts in all fields of astronomy unanimously agreed that "an instrument having the capability of the Advanced X-ray Astrophysics Facility (AXAF) is the highest priority for the 1980's."

This was a major boost for advocates of AXAF, considering that the Hubble Space Telescope had been ninth on a list of nine recommended major programs in the previous decade's report, *Astronomy and Astrophysics for the 1970's*. Doubly encouraging was the knowledge that the top program on the 1970s list, the Very Large Array (VLA) of radio telescopes, had been funded and built during that decade.

One disconcerting note in the midst of all the harmonious sounds coming from the scientific community was the result of a "costing exercise" for AXAF performed by the Marshall staff. In July of 1980, Jim Downey, the AXAF project manager at Marshall, presented the cost estimate for AXAF to Frank Martin and others at NASA Headquarters. Using the cost model developed for the Hubble Space Telescope, the staff had arrived at an estimated cost of $415 million in 1980 dollars, or about 80 percent of the cost projected for the Hubble Space Telescope at that time. This number raised eyebrows at headquarters, and alarmed Weisskopf and other scientists, who argued that Hubble was not a good model to follow. They were concerned that, if Hubble ran into problems, then, by association, so would AXAF. They proved to be right.

10

Jockeying for Position

BY THE TIME THE Astronomy Survey Committee published its report, the Einstein Observatory had ceased operations. Although astrophysicists would mine the rich Einstein database for years to come, Giacconi grew restless. He was now a Harvard professor and the recipient of many awards. He had an abiding interest in attacking scientific problems that would be illuminated by the Einstein Observatory, such as the origin and evolution of galaxy clusters, and he was the scientist in charge of AXAF, the National Science Foundation's highest-rated major new astronomy program for the coming decade.

I (WT) would meet with Giacconi during this time to work on articles reviewing the successes of Einstein. We discussed collaborating on technical and nontechnical books on X-ray astronomy. By their nature, these projects involved reflection on what had gone before and what lay ahead. It became clear that he was not unhappy with his career, as he had been in the Princeton days. Instead, the source of his discontent was another missing ingredient—action.

"I'm miscast," he said one day. "I'm not a professor. I'm a scientist and a manager."

AXAF was moving slowly through the system and Giacconi's experience told him that it would be years before he had anything significant to manage. His other consuming dream was to establish a national X-ray astronomy institute in the Harvard-Smithsonian setting. There again, progress was proceeding at a glacial pace, because such an institute would be predicated on the approval of AXAF. The major obstacle in the way of both these goals was the Hubble Space Telescope. NASA and Congress were chary about moving forward on AXAF until Hubble's schedule and budget were under control.

Then, suddenly, the obstacle became an opportunity. In the late spring of 1981, Giacconi was offered the position of director of the newly established Space Telescope Science Institute. The institute, which would be located on the campus of the Johns Hopkins University in Baltimore, Maryland, would be in charge of the science operations of the Hubble Space Telescope when it was launched.

Giacconi left Harvard and the AXAF program, he said, "to go where the work was." Later he said, "It was not exactly where I wanted to be. I came to Baltimore because I would have gone insane waiting for AXAF."

When Giacconi announced his decision to leave to the assembled high-energy group at Harvard-Smithsonian, everybody was shocked.

"There was stunned silence in the room," Steve Murray recalled. "We had no idea."

"The reaction at Marshall was mixed," Martin Weisskopf recalled. "On the one hand, I was concerned that Riccardo could not, and would not, be able to devote as much of his energy to AXAF. On the other hand, the team set up at SAO, including Harvey and Leon and others, was in place and working, and thus assumed to be a constant of the motion."

Once the shock started to wear off, speculation focused on who would be Giacconi's hand-picked successor. Harvey Tananbaum, who had been Giacconi's right-hand man since the Uhuru days and had coauthored the 1976 AXAF letter to NASA, seemed to be the front runner. Herb Gursky was also a strong candidate. He was senior to Tananbaum, had impeccable X-ray astronomy credentials that went all the way back to the original discovery flight in 1962, and had been a coauthor of the 1972 proposal for the original large X-ray telescope. Since moving to Harvard-Smithsonian in 1973, however, he had been associate director of the optical-infrared division, and had not been closely involved in X-ray astronomy. Yet another possibility was that a distinguished scientist from another institution would be brought in as the new group leader. Giacconi kept his own counsel.

Martin Zombeck reflected the general feeling of the Smithsonian group. "I didn't assume Harvey would take over. I didn't know who would."

At a dinner with Tananbaum and his wife, Rona, during that period, we asked him about his chances and the daunting prospect of succeeding Giacconi.

Harvey Tananbaum in 1984. (K. Tucker).

"I'm ready," he replied simply.

When Tananbaum is asked how he became interested in science, he will tell you that it was through baseball.

"I used to compute baseball averages and the changes in averages due to one more hit." The skill he developed in mathematics was enhanced during his high school years in Buffalo, New York, by a talented, honors program math teacher. He won a National Merit Scholarship to Yale University. He began with a major in mathematics but switched to physics. His choice of a graduate school was limited by an important criterion: it had to be in Boston, because that is where Rona, whom he was engaged to marry, had a job. He applied to and was accepted by MIT. He started in nuclear physics, then switched to the new field of X-ray astronomy and did a thesis on observations of Cygnus X-1.

"After graduating, I went around the corner to AS&E," he said. Over the next decade and a half he rose quickly through the ranks to become Giacconi's closest associate.

A few weeks after telling the group that he was leaving, Giacconi named his successor: Tananbaum.

The scientists in the high-energy group at Harvard-Smithsonian breathed a collective sigh of relief. If Giacconi had to leave, then Tananbaum was a popular and logical choice for a successor. It was felt that he would be a very effective proponent for AXAF within the NASA system and the scientific community.

"In some regard, Riccardo's move had its positive aspects," Weisskopf explained. "His outspoken disdain of bureaucracy and NASA—albeit usually justified—was beginning to rub too many open sores at NASA Headquarters. To that extent, the move did not hurt us at all. We ended up with the best of all possible worlds—an SAO that was easier to deal with and a powerful strong scientific advocate who was now the director of the Space Telescope Science Institute . . . Harvey played a critical part in the selling of AXAF to Congress—a role, quite frankly, that Riccardo could not have played."

When Tananbaum first took over the high-energy astrophysics group at Harvard-Smithsonian, AXAF was not his top priority. Getting more money to analyze the data from the Einstein Observatory was.

At NASA headquarters, George Newton, the head of advanced programs, worked to keep a modest amount of design dollars flowing to Marshall and Harvard-Smithsonian during a time when the astrophysics division was besieged by Hubble problems.

"George fought hard for those scarce funds to do that early work," Charlie Pellerin said, "to get the TMA [Technology Mirror Assembly] going. We were having overruns everywhere, and it was really hard to justify money for new programs."

Some NASA officials advocated that the agency should not even think about doing AXAF until Hubble was launched.

"I wasn't having any part of that," Pellerin asserted.

The Technology Mirror Assembly was a test article, the proof of a concept. The mirrors were comparable in size to Einstein's mirrors but were to be ground and polished to a smoothness five times that of Einstein's mirrors. This crucial program was begun in late 1982.

By mid-1983, with Einstein data-analysis funding in good shape, Tananbaum turned his attention to AXAF, which was progressing slowly. This was a danger. If a project stalls before it receives funding for the design and development phase—called a "new start" in NASA parlance—it risks being passed by another program or, worse, being abandoned by its advocates inside NASA.

"We were beginning to get frustrated," Tananbaum recalled, "so in June of 1983, George Clark [at MIT] and I got together and wrote letters to 240 of our closest friends in the astronomical community."

The letter gave the scientists information concerning AXAF and urged them to write an informational letter to Pellerin's boss at NASA, Burt Edelson, the Associate Administrator for Space Science. The thrust of their letters was that it was time to proceed with AXAF.

"Instead of worrying about a new start," Tananbaum explained, "they wanted specific actions."

Actions such as a request for proposal (RFP) to industry for preliminary design (Phase B) studies and an announcement of an opportunity (AO) for the scientists to propose scientific instruments for AXAF.

"Charlie Pellerin was initially skeptical about the impact of these letters, but they turned out to be very effective," Tananbaum said.

The letter campaign got AXAF moving forward again. Announcements requesting proposals for the prime contractors and for the scientific instruments went out later that summer. It was none too soon, because AXAF's position in the queue was being strongly challenged by another program, the Shuttle Infrared Telescope Facility (SIRTF), which proposed to put a large infrared telescope in orbit. SIRTF was behind AXAF in line and in the recommendation of the Astronomy Survey Committee report, but infrared astronomers felt that they had a good case for changing the order. The Infrared Astronomical Satellite (IRAS), launched in January of 1983, was sending back spectacular infrared images of cool clouds of dust and gas in our galaxy where stars are forming, of other galaxies where bursts of star formation were occurring, and of disks of possibly preplanetary material around nearby stars.

Late in the summer of 1983, a contingent of infrared astronomers led by Frank Low paid Pellerin a visit. Low is, in many ways, the infrared astronomy counterpart to X-ray astronomy's Giacconi. He had helped pioneer the field through breakthroughs in the design of infrared detectors, and he had made fundamental discoveries with his detectors.

"Frank showed up at my door one day with about five people. IRAS had been launched in January. It worked. In fact IRAS was startling. It was one of the greatest things we ever did."

By this time, the Hubble problems had been resolved, at least temporarily, so Pellerin and Newton were spending most of their time protecting money for AXAF.

"So, these guys showed up," Pellerin continued, "And they said, 'Look, X-ray astronomy has had Uhuru, HEAO A, Einstein'—I forget all the list of things. And they said, 'Infrared astronomy has nothing else but IRAS. We were promised by Frank Martin that if IRAS worked, SIRTF would go to the head of the queue.'"

Pellerin tried to tell them that he and others at NASA headquarters had been building the foundation and momentum for AXAF, so he could not support their request.

"They were really unhappy," he recalled. "They weren't hearing this."

The infrared astronomers were not the only petitioners.

"The UV [ultraviolet] people came in, and they were upset that FUSE [Far-Ultraviolet Space Explorer] wasn't happening, and all these people—all these guys—were coming in every week, a whole new group. Everyone was unhappy."

The infrared astronomers pushed to reconvene the Astronomy Survey Committee, which had given AXAF top priority, and have the committee members vote again. This raised the hackles of the X-ray astronomers.

"The X-ray people felt they had earned their position at the head of the queue," Tananbaum said. "Einstein had flown before IRAS, so it was logical that AXAF should fly before SIRTF."

George Field, the chair of the Astronomy Survey Committee, privately discussed reconvening the committee with the members.

"There was not much enthusiasm for this," Tananbaum said. "They had done a very careful and thorough job setting the priorities, and didn't see a clear path to changing their conclusions."

In October of 1984, Field wrote NASA a letter saying that the Astronomy Survey Committee stuck by its initial set of priorities. Later that month, Pellerin called Tananbaum and George Rieke, who was working closely with Low on SIRTF, to a meeting in Washington. They discussed the reasons why they were going to keep the order the same as before.

"What it boiled down to," said Pellerin, "was a question of readiness. AXAF was ready to move into the preliminary design phase [Phase B] and SIRTF was not. AXAF had been studied since 1977, the test mirror work was under way, spacecraft studies were under way, and the request for proposals for the scientific instruments and prime contractors had gone out."

"We thought we could do both eventually," Tananbaum said.

At the time, no one suspected how long "eventually" would be. AXAF's launch was still 15 years in the future and SIRTF was about 2 years behind AXAF.

"Had the SIRTF people known how long it would be, they might have been willing to fall on a sword," Tananbaum said, then added quickly: "Seriously, we have worked closely with SIRTF people, who have consistently supported AXAF and vice versa. We made a commitment in October of 1984 to work for the approval of both programs, and we have stuck by that."

Still, Pellerin was uneasy.

"I didn't know what to do about all this," he said. "It occurred to me that of all the people that had come to see me, no one had ever come in to talk about the kind of program we ought to have."

Pellerin did some research and made some inquiries about whether NASA had an overall plan.

"I looked around and I couldn't find one piece of anything to articulate a broader design. I started to tell people, particularly the infrared people, that we were having a stupid argument. To argue AXAF versus SIRTF is really an incredibly stupid thing because neither mission supplants the need for the other. Why don't we get on board something that gets us both? If you're really an astronomer, you need both. If you're not an astronomer, why are you talking to me?"

11

The Great Observatories

As 1984 wore on, Hubble's problems, never far below the surface, were bubbling up again. The fine-guidance sensors were late, and the computerized test system was not ready. In this context, Pellerin knew that it would be impossible to persuade Congress to move ahead on a major new program such as AXAF if astronomers were bickering among themselves about whether AXAF or SIRTF should go first.

"I thought, we really need some way to articulate the concept of the whole program and get everyone on board," he said.

He dropped into George Newton's office.

"George," he said, "I want you to go get the discipline chiefs together and get names of the dozen leading theoretical astrophysicists on the east coast . . . get good people representing all the fields . . . ask them to come to Goddard for a one-day meeting."

"What are we going to do?" Newton asked.

"We're going to color," Pellerin replied.

"What do you mean, we're going to color?"

"We're going to get crayons and magic markers and we're going to color."

To Pellerin's surprise, everyone Newton asked agreed to come.

"I realized," said Pellerin, "that I didn't know these people all that well. I had been in this job less than a year and these were big name people . . . I thought, I need to get somebody else who can run the meeting so I can think. Because I've got to be thinking ahead of these guys. Who should I get?"

Pellerin had recently finished reading *Cosmic Discovery*, by Martin

Harwit. In the book, Harwit, an astronomy professor at Cornell University, reviewed, classified, and analyzed 43 major discoveries in the history of astronomy, ranging from the discovery of planets in antiquity through the discovery of X-ray stars, pulsars, and quasars in the 1960s and gamma ray bursts in the 1970s. Full of tables, graphs, and recommendations, it was an ambitious and provocative book unabashedly aimed at policymakers in both the academic community and government. To judge from NASA's reaction to the book, Harwit hit his mark.

"*Cosmic Discovery* would turn out to be the centerpiece logic of the whole astrophysics division [of NASA] for years," Pellerin said.

What impressed Pellerin was Harwit's emphasis on the importance of new techniques that greatly expanded observational capabilities in more than one area. A hundredfold increase in the sensitivity, focusing ability, or the ability to measure photon energies precisely almost guaranteed new discoveries. The brief, discovery-rich history of X-ray astronomy was a powerful case in point.

"So, I called him up—I had never met him—and I said, 'I'm going to have this meeting, we're going to do this thing, and I'd like you to run the meeting.'"

Harwit agreed to help. Before the meeting he visited Pellerin at his Annapolis home to plan their strategy.

"Your job is to keep the meeting sort of moving in an orderly way," Pellerin told him. "Keep people on track. And my job is going to be trying to think what the hell we do next."

On the one hand, Pellerin seemed to be pursuing a strategy that Harwit had criticized in his book, namely centralized planning by a group of recognized experts, especially theorists, who he felt had undue influence over the field. Such planning, Harwit had written, "could easily choke progress and restrict the rate of discovery." On the other hand, Pellerin's objective was not to plan a new program. It was to reframe his existing program in a way that would appeal to the scientific community, then to NASA, and ultimately to Congress.

In their introductory comments at the meeting, Pellerin and Harwit pointed the scientists in the direction they wanted them to go. Pellerin asked them to list the top 10 most interesting problems in astrophysics. Once the group agreed on the list of important problems, Pellerin passed out large sheets of paper, magic markers, and crayons and asked them to draw pictures that illustrated the problems.

"I want you to draw pictures of something that shows what this looks like," Pellerin told them. "For every image you make, I want it to make the argument for getting complete physical understanding and the need for measurements across all the wavelength regimes."

The theorists' final list at Pellerin's meeting covered 10 issues that they considered of paramount importance, with the types of telescopes needed to make significant progress:

What is the life cycle of stars? Radio and infrared telescopes are needed to study the formative phase of stars, optical for the middle phase, and X-ray for the end phase.

What are black holes? Do they really exist? Optical observations would measure the effects on the orbits of stars near black holes, while X-ray and gamma ray observations would study the gas close to the event horizon.

How do quasars generate so much energy? To answer this question the same things are needed as for the study of black holes, except infrared observations are needed if the quasar is shrouded in dusty gas, and radio observations are also useful to study orbiting material and the high-energy jets commonly seen around quasars.

How are magnetic fields produced in the universe? Radio, optical, and X-ray telescopes can study the magnetized jets; X-ray telescopes can study magnetized gas in flares on the surface of stars; and X-ray and gamma ray telescopes can study extremely strong magnetic fields on neutron stars.

What is the nature of the dark matter? Radio and optical telescopes are useful for studying the effects of dark matter on orbiting stars and gas clouds in galaxies. X-ray telescopes can measure the amount of dark matter in galaxy clusters, and infrared telescopes can detect small planet-like objects called brown dwarfs, if they are the dark matter.

How did galaxies and clusters of galaxies form and evolve? Sensitive infrared and optical telescopes are needed to study the most distant galaxies, while large X-ray telescopes are needed to see the hot gas in the most distant galaxy clusters. Microwave observations will provide essential details on the state of the universe before galaxies formed.

What is the geometry of the universe? Is it finite or infinite? An understanding of the geometry, size, and fate of the universe is tied to an understanding of the nature of dark matter and the formation of galaxies and galaxy clusters.

Did the universe once contain equal amounts of matter and antimatter? Does it still? Antimatter and matter annihilate on contact with an explosion of gamma rays. No antimatter galaxies are known to exist today, but if they did exist in the distant past, a gamma ray telescope should be able to detect radiation emitted when antimatter ejected from such a galaxy comes into contact with ordinary matter from a galaxy such as ours.

How are planetary systems formed? How many stars have planets? How many might be habitable? An optical telescope such as Hubble can find the planets, and an infrared telescope such as SIRTF can reveal the details of the dusty gas from which the planets formed, and search for telltale molecules in planetary atmospheres that could indicate whether a planet is habitable.

Is life on Earth unique? The surveys described above will give us a firmer foundation for making estimates, but we will probably have to rely on contact via radio telescopes to know for sure.

Pellerin collected the images and announced, "It looks like we need four major space missions."

The astronomers, with George Field taking the lead, agreed. Of course, they were all theorists, not observers pushing their particular mission, and an outside observer might be forgiven for believing the fix was on, that Pellerin had already made up his mind and was seeking the imprimatur of the elite. At any rate, Pellerin felt that the list was a significant step forward.

"What we did is to put this thing together and get it anchored around four main observatories—Hubble, GRO [Gamma Ray Observatory], AXAF, and SIRTF," he explained. "Then everything else could flow in under it to support it . . . My basic strategy was that if we could show something that is logical, we could start people thinking that we needed all the missions together to get the job done. Then I could stop the infighting, get people to really believe that we had a sequence and the best way to get it was to support AXAF. That was the idea."

Pellerin had the vision. Now he needed a name to complete the package. But he was stumped until George Field came down to Washington from Harvard for a meeting. He stopped by to see Pellerin and discuss the progress on selling AXAF.

"George, I've got it," Pellerin told Field enthusiastically. "We have these four observatories and we have this set of things that underpin it, supporting missions. It's such a great program, but I can't think of a name for it."

"What do you mean?"

"It's so great, but I'm completely lost," bemoaned Pellerin. "I can't think of a name."

"Why don't you call it the Great Observatories?"

Pellerin knew immediately that he had a name for his program. He asked Harwit to put together a booklet that articulated the Great Observatories concept based on the words and the pictures that the astrophysicists had produced.

"What I want is a thing with cartoons on one side," he told Harwit. "I want this to look like something that people won't be afraid of."

"The obstacle that I had on the Hill," he explained, "was that if you went in there and said 'astrophysics,' then people froze up, thinking 'this guy's going to discover that I failed a course in chemistry or something.' So most people on the Hill have different skills that I learned to admire, but they are not interested in the calculation of stellar atmospheres."

Pellerin, Harwit, and Valerie Neal of the Essex Corporation quickly put together a 50-page cartoon book entitled *The Great Observatories for Space Astrophysics.* They distributed it widely among the opinion-makers on Capitol Hill, within NASA, in the scientific community, to schools, everywhere.

"We were dropping it out of airplanes, basically," Pellerin joked. "The whole idea was to put AXAF in a context so that AXAF was the key to everything. That is, if we could get everyone to back AXAF, then we could get everything."

In this situation, everything meant AXAF and SIRTF. Hubble was already being built, and the Gamma Ray Observatory had been funded through Frank Martin's efforts before he left NASA.

"The idea was that we offered a completion point," Pellerin said. "It's not an infinite program, but we do need these four things, and this will change the face of our understanding of our place in the universe."

Pellerin was not just counting on the intellectual acceptance of his program. He knew he had to connect at a deeper level.

"I was aware that NASA was supported by several of the primary myths of our time," he said. "One was the hero myth. That was the one that was keeping the astronauts moving. The second was the myth of expansion—the notion that we are somehow here to conquer whatever is here, and having conquered the west and conquered the world, we are going to conquer space—so we played into the idea of the Americans being the first to do this."

12

The Belmont Retreat

THE GREAT OBSERVATORIES concept gave the AXAF program a big boost. Contracts were let to TRW and Lockheed for observatory definition (Phase B) studies. But the Hubble Space Telescope continued to have problems. The testing of the myriad components of the spacecraft was moving ahead slowly and new troubles with the software seemed to crop up daily.

When Hubble coughed, the other astrophysics and space science programs at NASA got pneumonia as more money flowed out of their budgets into Hubble's. During this period, the space science programs in the Office of Space Science and Applications were divided into four broad categories: astronomy, which refers to everything outside the solar system; planetary exploration; solar-terrestrial science, which refers to studies of the sun and its interaction with Earth's magnetic field and upper atmosphere; and Earth sciences, which refers to the study of Earth from space. Although all these disciplines are technically space science, the last two categories are commonly referred to as space science, as distinct from astronomy and planetary science.

"They [the planetary and space scientists] hated Hubble," Pellerin recalled. "The planetary scientists dreamed of a return to the glory days of the Mars Viking program, and the space scientists—scientists who studied the upper atmosphere and space environment of the Earth and the Earth from space—felt that their programs were being neglected. They wanted a big program of their own, like TOPEX [Topography of the Ocean Experiment] and ISTP [International Solar-Terrestrial Physics satellite].

Even though they were hurting because of Hubble, both the planetary and space scientists knew how to play the political game.

"They would invite Burt Edelson [the Associate Administrator for Astrophysics and Space Science at NASA] out for a gala party at times that just happened to be just before major budgetary decisions," Pellerin observed. "And they were able to stack the committees within NASA."

"They were always afraid of astrophysics," Pellerin said, "because they felt that any outside group would support astrophysics over space science research."

Pellerin knew, however, that the support of an outside scientific group, however prestigious, would not be enough to promote AXAF to "new start" status, a necessary step before funds are committed to build a space observatory.

"The upper echelons of NASA are not science driven," he explained. "There must be a political signal."

One came in the summer of 1984. George Keyworth, President Reagan's science advisor, wrote a letter to NASA Administrator James Beggs. Dated June 4, 1984, it was written on stationery with an austere letterhead stating simply: The White House, Washington. In it, Keyworth expressed concern about the declining fraction of NASA's budget devoted to solar-terrestrial and Earth sciences, and urged more support for studies of the sun and the Earth from space. He concluded by writing that "we must find effective ways to move this discipline forward in balance with the areas of astronomy and planetary exploration."

Keyworth's Office of Science and Technology Policy followed 2 months later with a report entitled "Funding Trends in NASA's Space Science Program." It was an 18-page document with graphs and pie charts conveying essentially the same message as Keyworth's letter: astronomy's slice of the pie has gotten twice as large as any other element in the Office of Space Science and Applications budget, and it was time to restore some balance.

Pellerin immediately fired off a rebuttal giving his analysis of the same facts and figures. He pointed out that the assignment of funds to the various disciplines was somewhat arbitrary, since some of the programs had overlapping missions—for example, Hubble does some planetary science—and he argued that the relative percentage of the astronomy budget was scheduled to decline in the existing plans, so that there was no need for a further revision of the balance between astronomy and solar-terrestrial physics.

His points were well taken, but Pellerin knew that the damage had been done. The political signal had been sent.

"I was frustrated. I had the number-one project. I never doubted that it was what we should do."

He stressed the need to get AXAF ready so that it could be launched shortly after Hubble, giving astronomers the ability to observe the cosmos with optical and X-ray telescopes of comparable power. This was part of his strategy to counter the political clout of the space scientists. His real worry now was that AXAF would be dropped from NASA's agenda altogether.

"They can say 'no' to you at five different levels. If the Associate Administrator of Space Science says 'no', it's over. If the administrators as a group say 'no' when they make up their budget for NASA, it's over. If the Administrator says 'no' before sending his budget to OMB [Office of Management and Budget], it's over. If the OMB says 'no', it's over. And if they say 'no' on the Hill, it's over. The way it works is that people get rid of things. I had to make AXAF look so strong and so logical that people would be afraid to oppose it. I had to create fear of a lost opportunity."

One might have thought that the National Academy of Science Astronomy Survey Committee or the Field Committee report that had listed AXAF as the number-one priority for the 1980s was sufficiently impressive to create such fear. Not so, said Pellerin.

"After 2 years it became stale," he explained. "It was necessary to get us going, but not sufficient to keep us going."

Pellerin expanded his advisory group to include more scientific luminaries to give it a high profile, named it the Astrophysics Council with Martin Harwit as chair, and scheduled bimonthly meetings to provide opportunities for the scientists to drop in on members of Congress and their staff.

According to Pellerin, certain NASA officials wanted to abolish the Astrophysics Council. They were ultimately afraid to shut it down, Pellerin said, "because I had energetic, powerful people on it. Nobel laureates and people like Riccardo, and other space astrophysicists. They were brawlers, scrappers, and hell-raisers." Pellerin's strategy began to have an effect. Through the efforts of the Astrophysics Council, the Great Observatories concept became a subject of conversation among planners in high places in NASA and on Capitol Hill.

"We were rolling along," Pellerin said. "Meanwhile, Burt Edelson is going nuts. He wanted to promote materials research, but most of that kind of work is done in industry . . . He liked things that are doable. Astrophysics was expensive and risky, but it was what everyone was talking about on the Hill."

In the spring of 1985, Edelson needed to lay out the "new start" priorities of the Office of Space Science and Applications for the next decade.

"This was a problem for Burt," Pellerin said. "He hated to make decisions, and there was never enough money to do all the good projects that were proposed."

Edelson decided to call a retreat for managers from NASA Headquarters and the field centers at a resort in Belmont, Maryland. When Pellerin arrived, he did not like what he saw.

"I was hopelessly outnumbered," he recalled. "I looked around the room and I didn't see a lot of people that had astrophysics as a high priority . . . everyone was mad about Hubble. It felt like Custer's last stand, and they were going to set the priorities for the next decade."

The managers did not get to a discussion of priorities immediately. First, they discussed the management and operation of the Space Station.

"I never did get on board with that one," Pellerin said. "I couldn't see the relevance."

Nor did they discuss new starts in space science the second day. Despite the efforts of a professional facilitator, they kept veering away from the stated goal of the retreat.

"It was completely off the wall," Pellerin said. "The facilitator was getting worried. He was aware of the gravity of the situation, and was concerned by the chaos."

As a means of bringing some order and focus to the discussion, the facilitator proposed a straw vote. He emphasized that the vote was a common technique to get people talking, and carried no weight in the final decision.

"This made me really nervous," Pellerin said. "Every neuron in my body was telling me that this was really dangerous, because Burt does not want to deal with this, and the first excuse he gets to not have this discussion continue, he's going to take it."

Pellerin figured that Edelson would turn the straw vote into the real vote, and it would stand as the definitive statement of this administration on "new starts" for the next 10 years. As he watched the facilitator prepare for the coming vote, he realized that it could very well determine the fate of AXAF.

The group of 15 ended up with 13 new possible start projects. The facilitator gave each person 2 votes per item, or 26 votes total, to be distributed among the various projects. Sam Keller, the deputy associate administrator, was first to vote.

"Sam was a powerful, forceful person who had been around a long time," Pellerin said. "He was a lot like Lyndon Johnson in his political ability to get things done in the system. He had done a lot to keep up support for Hubble on the Hill. He sold the overruns to Congress without a flap. I figured that he would definitely be a model for a lot of people in the room as to how to vote, because the tension in the room was very high. After years of work, we were starting the process of deciding which projects would make the 'new start' list for the next 10 years."

Keller distributed his votes among the various projects, giving the most votes, 4, to Cassini, a planetary probe to explore Saturn, its rings, and its moons. He gave AXAF 1 vote.

"Burt was grinning happily at this point. Things were going the way he wanted."

Four or five other participants voted. TOPEX, the project to explore the ocean through remote sensing, emerged as the leader. AXAF, getting at most 1 vote per person, was trailing badly. Then it was Pellerin's turn. He recalled the moment:

"How do you vote, Charlie?" the facilitator asked.

"Twenty-six votes for AXAF," Pellerin replied.

"You can't do that!" Edelson protested.

"I thought we could apportion the votes however we wanted," Pellerin maintained.

"Yeah, that's right," the facilitator admitted. "But no one has ever done this before."

"I thought we were supposed to vote according to the way we think things should be. This is way I think things should be," Pellerin countered.

"Burt was red-faced," Pellerin recalled, "And Sam, he just buried his head in his hands. I knew what he was thinking. He was thinking, 'Oh, my god, Charlie's done it again.' There I was, the frog in the punch bowl, you know."

The facilitator let Pellerin's vote stand, and continued around the room. None of the other participants opted to use Pellerin's strategy.

"No one even imagined doing that," Pellerin said.

In the end, TOPEX was clearly number one, with AXAF, CRAF (Comet Rendezvous and Flyby), and ISTP grouped together as second best.

After the vote, the facilitator called for a break for lunch, after which the group was supposed to return to discuss the results, and then vote again. During the lunch break, Edelson declared the retreat over.

13

Progress and Setbacks

CHARLIE PELLERIN'S unorthodox tactics at the Belmont retreat had kept AXAF alive, but just barely. A month later, NASA's Space and Earth Science Advisory Committee (SESAC) met to vote on priorities for major new missions, or "new starts," for the fiscal year 1988 budget. Pellerin was worried. Astrophysics had only 9 members on the 29-member committee, which was chaired by Lou Lanzerotti, a space scientist from Bell Telephone Laboratories.

"Lou Lanzerotti seemed to have a deep animosity toward the astrophysics program," Pellerin said, "probably because he felt that we had stolen the Explorer program away from space science, and spent all the money on IRAS [the Infrared Astronomy Satellite]. I thought he was very biased and unfair in the way he ran the committee and told him so, which probably didn't endear me to him, but that's my way."

Pellerin recognized that TOPEX (the Ocean Topography Experiment) would be the committee's first choice. He also knew that AXAF was unlikely to be the second choice, and therein the danger lay.

"I was a better than average infighter, so if we could get out of there without an agreement, I could do something. But if they steamrolled this thing with a CRAF [Comet Rendezvous and Flyby], ISTP (International Solar-Terrestrial Physics satellite), AXAF sequence, AXAF would stop work. Someone would take my advanced program money out; they would stop the procurements. It is the easy way to stop a program."

Years ago, when we were undergraduates at the University of Oklahoma, we would drive through the hills of southeastern Oklahoma in an old Nash Rambler. One winter solstice, we were overtaken by an ice storm, which began coating the road with an ever thickening film of ice.

The hills in that part of the state are not large, but there are many of them. We soon learned that maintaining steady forward progress, however slowly, was our best strategy. Once the wheels started to spin, one of us (WT) would jump out and push to get us over a particularly slick spot before we started to slide backward.

In 1985, AXAF was moving into dangerous territory just like that little Nash Rambler. If anything stopped its forward progress, however slow, it would start to slip backward, with possibly dire consequences. Pellerin understood this clearly.

"When a project gets big," he explained, "it's hard to make a unilateral action because there's lots of political forces that inquire into the logic and so forth. But if you want to shut me down, all you have to do is take the list—CRAF, ISTP, AXAF—and the logic would be that we're going to get at best one every year, one every 2 years, so therefore it's really dumb to spend that money on AXAF. I would be immobilized. I would lose the civil service support, and I can't go out and appeal to the administrator for $2 million dollars. I couldn't go around my management for $2 million. They would never hear of it."

Fortunately, there were people around willing to do the pushing necessary to keep AXAF moving, to move it over the slick spots. There was Pellerin, not just "a better than average infighter," but as MIT's Claude Canizares rated him, "the consummate infighter." There was Canizares himself, a member of the SESAC committee, whose solid scientific credentials and diplomatic style won him friends and influence on the committee.

"I took the point of view that politics is the art of the possible," Canizares told us. "I tried to build relationships with the committee members, and to work toward a compromise wherein we would endorse TOPEX, but keep the number-two position uncertain."

"This was a place where we could have had a complete and total loss," Pellerin emphasized. "But Claude was so clever, so articulate, so brilliant, that he managed to keep it uncertain. With that uncertainty, I could proceed and do the bureaucratic part better."

Another key "pusher" was George Field, then director of the Harvard-Smithsonian Center for Astrophysics, and chair of the National Academy of Sciences Astronomy Survey Committee that had listed AXAF as the top priority for all new astronomy projects for the 1980s. Field was a willing and articulate spokesman for AXAF in the upper echelons of NASA.

Riccardo Giacconi was now director of the Space Telescope Science Institute and had his hands full trying to cope with problems not of his own making with the Hubble Space Telescope. Nevertheless, he found time to remain involved with AXAF through the Astrophysics Council and other committees. He used his growing access to the high and mighty in Washington to put in a word on behalf of AXAF whenever possible, even during a luncheon with President Reagan.

At this luncheon, which was also attended by Edward Teller, George Keyworth, then science advisor to Reagan, and White House chief of staff Donald Regan, among others, Giacconi talked about the Hubble Space Telescope and NASA's Great Observatories program. He drew a comparison to the Renaissance period of the sixteenth and seventeenth centuries, which saw increased exploration by Columbus, Magellan, Cartier, Drake, Cortez, and Raleigh, and the flowering of inquisitive spirits such as Copernicus, Galileo, Kepler, and Newton. In a similar way, he stressed, we must embark on an exploration of space with our minds as well as our bodies. Teller expressed his conviction that astrophysics was perhaps one of the most interesting fields in physical science and, according to Giacconi, spoke eloquently in support of AXAF, stating that "the discovery of a black hole would be one of the great discoveries of the twentieth century." Giacconi commented that the president seemed generally sympathetic to the subject. It is impossible to say whether this type of meeting helped AXAF's cause, but it certainly could not have hurt it.

Martin Weisskopf also played an important and supportive role. As the project scientist at the Marshall Space Flight Center, his grasp of the technical details and the scientific import of AXAF made him an especially effective advocate for AXAF within NASA and with Congressional staffers.

Harvey Tananbaum was everywhere, scanning the scientific and political air space like radar, recording every blip in the program and checking out each and every one to make sure that AXAF stayed on track. This could mean a trip to Danbury, Connecticut, to check on the progress of the Technology Mirror Assembly, a meeting with Weisskopf and the rest of the project team at the Marshall Space Flight Center, attendance at an Astrophysics Council meeting in Washington, the preparation of fact sheets and brochures to explain the importance of AXAF to NASA officials or Congressional staffers, a meeting with these officials and staffers, or the organization of a nationwide network of scientists who could be called upon to write letters or visit NASA officials or members of Congress.

"He became a major force," Pellerin acknowledged.

During this critical period a number of people played important roles in keeping AXAF moving. But the most important factor was not the work of any one individual, however dedicated and energetic. It was work in the laboratory by scientists and engineers associated with AXAF that showed that a large X-ray observatory was feasible, and research by the larger scientific community that demonstrated that X-ray observations were vital to understanding some of the most pressing scientific questions of the day.

The Technology Mirror Assembly program, conceived by Leon van Speybroeck and Martin Zombeck of the Harvard-Smithsonian Center for Astrophysics, was a way to demonstrate that the technology existed for grinding and polishing mirrors to the exquisite precision planned for AXAF. Specifically, the requirement was that more than half the X-rays that the telescope received from a distant X-ray source would be focused onto an area 1/100th the size of any previous X-ray telescope. Stated another way, the mirrors had to be able to resolve sources that were less than half an arc second in diameter, or more than 3,000 times smaller than the angular diameter of the moon. This requirement is equivalent to the ability to read the letters on a stop sign at a distance of 12 miles (20 kilometers). It meant grinding and polishing the mirrors to such a smoothness that any lumps would be less than about 6 atoms high. For comparison, the width of a human hair is equivalent to about 1,000,000 atoms set side by side.

Many were skeptical that this could be done, but not the Harvard-Smithsonian X-ray astronomy group.

"We were always extremely confident we could do mirrors well, if they would only listen to Leon," recalled Dan Schwartz of Harvard-Smithsonian, who directed the testing of the Technology Mirror Assembly.

Indeed, when the tests were performed at the X-ray Calibration Facility in Huntsville, Alabama, they were in excellent agreement with the predictions by van Speybroeck and Paul Reid of the Perkin-Elmer Corporation.

"The tests were impressive," Schwartz said. "They showed that we could achieve extremely high accuracy in our tests, and they proved that we could make the half arc second mirror needed for AXAF."

Van Speybroeck, who can be brutally honest, especially when it concerns his own work, was not so sanguine.

"The NASA reviewers were favorably impressed by the technology," he said. "But they were concerned with scale-up issues. The AXAF mirrors would be much larger and more difficult to handle. We would have to use cranes to move them around. They were right to be concerned. It was going to be difficult."

On the whole, though, the reviewers gave the AXAF team high marks.

"It [the review] was brilliantly accomplished", Pellerin said. "George Newton [deputy director of astrophysics under Pellerin], the mission support team, and the Marshall team did a great job. We were checking boxes off really carefully . . I knew there were people around who didn't give a damn whether we went to Pluto or Mars or whatever, but they cared a lot about whether we had checked the boxes off."

While the AXAF team was checking off boxes, and pushing ahead on the technical and political fronts, X-ray astronomers were digging deep into the rich lode of the Einstein Observatory database. Among the many intriguing discoveries that emerged from this research was the demonstration that X-ray observations were one of the best means for tracking down the elusive dark matter in the universe.

By the middle of the 1980s, the mystery of dark matter was rapidly becoming one of the major unsolved problems in astronomy. Strong evidence had accumulated from radio and optical observations that galaxies are embedded in massive clouds of dark matter. The inferred amount of dark matter was mind-boggling. Fifty to 80 percent of every galaxy seemed to be in the form of dark matter. Galaxy clusters, the largest gravitationally bound objects in the universe, appeared to contain an even larger percentage of dark matter. If this was true, then dark matter represents the most prevalent form of matter in the universe and astronomers did not know what it was.

One possibility that had been discussed in the late 1960s was that the dark matter might be in the form of very hot gas. Such gas would radiate primarily in the X-ray band while producing an undetectable amount of radio or optical radiation, so it would show up once X-ray telescopes became sensitive enough to detect it.

Indeed, Uhuru did observe X-rays from clusters of galaxies in 1972. But other X-ray satellites and small X-ray telescopes aboard rockets in the late 1970s established that the dark matter is not in the form of hot gas. They did, however, demonstrate that a focusing X-ray telescope could be a powerful tool for tracing dark matter, because the temperature of a gas is a measure of the average speed of the atomic particles in the gas, and this speed can in turn be related to the gravity necessary to hold the particles in a galaxy. It is the same principle rocket scientists use to calculate the final speed a rocket must attain to escape from the Earth's gravity. For the Earth, the escape speed is 25,000 miles per hour. For the moon, it is only about 5,000 miles per hour, which explains why the moon doesn't have an atmosphere worth talking about. At typical lunar temperatures, many of the gas particles have speeds in excess of the escape speed, so any gas there has long since evaporated into interplanetary space. Whether a planet can hold onto a cloud of gas depends on its temperature and its gravity, which depends on its mass and size.

With the Einstein Observatory, astrophysicists could make fairly accurate estimates of the temperature and size of the hot gas clouds around galaxies or in galaxy clusters. Using this information, they could estimate the gravity and hence the mass that the galaxy or cluster needed to hold the gas in place.

One of the first demonstrations of the power of this method was made by Daniel Fabricant, Paul Gorenstein, and Myron Lecar of the Harvard-Smithsonian Center for Astrophysics. They showed that the giant elliptically shaped galaxy M87 was one the most massive galaxies ever observed. Roughly 5 percent of its mass can be accounted for by stars. Surprisingly, the 30-million-degree gas cloud in the galaxy contains as much mass as all the stars. Even more surprisingly, 90 percent of the total mass of the galaxy must be in the form of dark matter to keep the gas from escaping.

The husband-wife team of William Forman and Christine Jones of Harvard-Smithsonian, Andrew Fabian of Cambridge University, Canizares, and others quickly applied this technique to other galaxies and clusters. It was soon apparent that X-ray observations would be essential in the quest to understand the nature of dark matter. Sandra Faber of the University of California at Santa Cruz, an optical astronomer who had done important research on dark matter, acknowledged the importance of X-ray observations in her keynote speech at the International Astronomical Union symposium on dark matter at Princeton University in June 1985, saying that "AXAF will be crucial for measuring accurately the amount of dark matter in clusters of galaxies".

It seemed that astronomers of every stripe were convinced that AXAF was needed. It seemed that AXAF was ready to proceed technically. The X-ray astronomers had bright, diligent, well-organized leaders like Tananbaum and an effective street-smart insider in Pellerin. Yet AXAF was not going anywhere. It was just spinning its wheels, and it was all these brilliant, hard-working people could do to keep it from slipping backward. What was wrong with this picture? Why could AXAF not get approval for a "new start"?

In the mid-1980s, several factors held AXAF back. The first—stiff competition—was not unique to the times. There are always more excellent, highly deserving projects than can be funded, so it is always tough, very tough, to get a "new start" for a large project. Then there was the Hubble factor. AXAF was the next large astrophysics project to go through the system after the Hubble Space Telescope.

"It's like following a big elephant down the road," Canizares explained. "If you follow it too closely, lots of things can happen, and not a lot of them are good . . . I remember one time we had lunch with Burt Edelson at the Cosmos Club, and he asked, 'Why does AXAF have to look so much like Hubble. Can't you make it a cube, or at least a rectangle?'"

The problems with Hubble were symptomatic of much larger problems within NASA that also affected AXAF. The planned Space Station was coming under increasing criticism by scientists, who feared that it would suck up all the money for scientific research. The military had designs on the Space Station and the Space Shuttle program. NASA Administrator James Beggs, a strong and vocal defender of his turf, convinced President Reagan that the shuttle program and the Space Station would be a commercial success if NASA retained control and increased the number of shuttle flights.

In December of 1985, Beggs was indicted by a federal grand jury on charges related to a government contract at General Dynamics, where Beggs had been a vice president before becoming NASA Administrator.

"DOD [Department of Defense] got Beggs," Pellerin said, echoing a common belief. "He was too big."

While under indictment, Beggs was put on leave from his job and replaced by William Graham, who had been hand-picked by influential supporters of the military. Ironically, any ideas of a "military takeover" of the Space Shuttle and Space Station programs were soon overwhelmed by a disaster that demonstrated the failings of the NASA management—the tragic explosion of the Challenger Space Shuttle on a cold January morning.

The shuttle fleet was grounded, a blue-ribbon investigative panel was formed, and sweeping changes were made in NASA's administration. Beggs resigned, though the charges against him were expunged. President Reagan, in an effort to restore confidence in NASA, passed over Graham and appointed James Fletcher as the Administrator of NASA. Fletcher had been Administrator in the 1970s, and was instrumental in starting the Space Shuttle program. He had also approved the funds for the preliminary study of AXAF, and was well aware of the successes of the Einstein Observatory.

Having Fletcher as Administrator was a definite boost for AXAF, Pellerin felt. "We had a bit of a relationship from when he had been administrator before. I began to cultivate it. He was a scientist and I was a scientist and so we had this 'we're scientists talking' type of thing going for us."

Of course Fletcher and NASA had a lot more things to worry about than AXAF in 1986, so Fletcher's presence made little difference. At its June meeting, the SESAC committee put AXAF third behind the TOPEX program, now renamed Global Geo-spacescience Satellite, and CRAF. Only the Global Geo-spacescience Satellite was recommended for a "new start."

"I was furious," Pellerin remembered. "I felt that the whole process was rigged." Upon reflection, he added, "Basically what happened was that we had Hubble problems, so we were third in this list and it couldn't be stopped. But at least we were in the list."

Pellerin tried another tactic. He created a "pathfinder" program, according to which the AXAF team would grind, polish, and coat one pair of mirrors, and study but not build the scientific instruments. This way they could keep the program moving, demonstrate that the largest mirrors could be built, and get realistic estimates of the final cost of the complete observatory. Although Burt Edelson was lukewarm about the idea, Fletcher supported it and sent it up to the Office of Management and Budget (OMB), which turned it down.

"Those days were the most stressful days of my entire professional life," Pellerin said. "I felt misunderstood, I felt rebuffed at every turn. I would have quit if it had not been for two things. I still had this sense that this was what I was intended to do, to make this program go, and that no one could do it as well as I could."

III

THE MIRROR CHALLENGE

14

The Challenge Is Set

IN EARLY 1987, Charlie Pellerin's unquenchable enthusiasm was about to be quenched by his inability to move the AXAF program forward.

"If a program the size of AXAF hangs around too long and doesn't move forward," he explains, "people begin to assume that something's wrong with it, and you're dead."

The tenth anniversary of the approval of funds for the feasibility study of AXAF was approaching. It had been around a long time.

"Then, a minor miracle occurred," Pellerin related. "Burt Edelson resigned and Lennard Fisk became the Associate Administrator for Space Science."

Pellerin and Fisk had occupied offices next to each other in the days when they were both at the Goddard Space Flight Center, working "back there where the phone never rang," as Pellerin put it. As they worked together at NASA Headquarters, this long-standing relationship grew into a mutual admiration society.

"Len Fisk was one of the most incredible people I have ever been involved with," Pellerin said. "He always acted from a place of principle."

For his part, Fisk said that "one of the pleasures of this whole effort was having Pellerin to work with . . . I valued him enormously."

Fisk's appointment was part of the shakeup in NASA management that followed the Challenger tragedy. When we talked to him several years afterward, he explained why he left a comfortable job as a dean at the University of New Hampshire to jump into the turbulent, troubled waters of NASA.

"I think a lot of us at that time thought the space program was in lots of trouble," he said. "There was a kind of an obligation to go and be part of the fix."

There was a lot to fix.

"We were only a year away from Challenger. Nothing was flying. The whole program was on the ground."

Fisk already knew about AXAF, because, as he joked, "there was no committee that I had not served on." Nevertheless, Harvey Tananbaum, who worked on the assumption that you could never know enough about AXAF, went with Jonathan Grindlay, a Harvard professor, to visit Fisk in New Hampshire, and to supply him with the latest information on AXAF's progress. Fisk was favorably impressed. He realized that AXAF might be a way to resuscitate a moribund space program.

"Len Fisk could always find the third way of looking at a problem, when other people might be stuck with one or two ways," Pellerin remarked.

In the case of AXAF this third way consisted of reframing the context of the problem from an intra-agency political battle over how to divide up a diminishing budget pie to a geopolitical issue.

"The Russians were strutting their stuff," Fisk said, referring to the space station Mir and the Russian launch rate of almost one satellite per week in the year following the loss of the Challenger and its crew. "They were claiming that they were going to go to Mars, and they were going to have this big bold program, and we couldn't even get anything off the ground. There was a real motivation to do something that stated that the U.S. was going to be a leader in space science. We wanted to make a statement. We wanted a new start for something significant."

At that point the choice was AXAF versus CRAF (Comet Rendezvous and Flyby).

"CRAF was not a bestseller," Fisk said."I personally like comets. Comets are very interesting, but somehow it wasn't selling."

It was here that the groundwork on AXAF laid by Tananbaum, Pellerin, and others paid off. NASA Administrator James Fletcher, congressional staffers, and science policymakers in the White House such as Bill Graham, who had moved on to become the head of the White House Office of Science and Technology Policy, and George Keyworth, who was the

President's science advisor, all knew about AXAF and its potential to make major scientific discoveries.

"AXAF fulfilled that desire to have a leadership program—to make a statement that we were going to get back into space science and be at the top of the heap again," Fisk said. "Here was something the U.S. could do uniquely."

Even so, Fisk was not under the illusion that AXAF would sail through the approval process. "There was the Hubble legacy," he explained. "The Hubble had not flown, the hype was still there, but there were a lot of bad memories, and the idea of starting another one before you've finished the first one was not a thrill."

Tananbaum tried to counter this impression in the booklets, fact sheets, and briefing materials his team prepared. He emphasized important differences between Hubble and AXAF. Hubble had expensive fine-guidance sensors that AXAF did not have. X-rays are much less plentiful than optical photons from a typical source, so AXAF could get by with a much slower rate of accumulating data than Hubble needed. Finally, unlike Hubble, the AXAF program had a group of scientists and engineers from several institutions that had years of experience building X-ray mirrors and doing space projects.

Fisk decided that it was time to take action on AXAF, for better or worse.

"Your project comes up through the system," he explained. "The science community decides what it wants to do and it sort of works its way up through the system. Then you come up to bat and if you don't get on base, somebody says, 'Why are you still here?' It was clear that AXAF had been around for a while and it was time to either get it through or rethink what was going to happen. There was a need to make it into a real program."

In July of 1987, Fisk, Pellerin, Tananbaum, and other scientists made a presentation to Fletcher. Two months later, in September, he announced his decision: funds for a new start for the AXAF program would be included in the budget that NASA submitted to the Office of Management and Budget (OMB).

At the Harvard-Smithsonian Center for Astrophysics, the X-ray astronomy group gathered in Tananbaum's office to watch him cut a cake decorated with icing in the shape of an X-ray telescope. He congratulated his group and paid tribute to Riccardo Giacconi's vision. (The last time the group had assembled in that office was 5 years earlier to hear Giacconi's announcement that he was leaving to become director of the Space Telescope Science Institute.) Fred Whipple, an 80-year-old senior statesman of astronomy who had brought the Smithsonian Astrophysical Observatory into the big time by moving it from Washington to the Harvard University campus in the Sputnik era, heard the commotion and walked down the hall to join the celebration. Tananbaum warned of the hazards of the road ahead as AXAF made its way through the budgetary process and emphasized the need for teamwork in the hard work that lay before them.

"The key was teamwork." Tananbaum recalled. "We set up a good coordination team with NASA, scientists, and industry. We kept each other informed. We had one story. We shared contacts and information. And we had a lot of facts to support our case."

In a few months, all these resources would be needed to save the project. After NASA submits its budget, it is reviewed by the Office of Management and Budget (OMB), one of the most powerful and feared offices in the administration. If that office rejects your project, it is not dead but it is sick. You must then appeal to the President, who consults the appropriate advisors and makes a thumbs-up or thumbs-down decision, which is final.

"Usually you are told in September what OMB will do," Pellerin told us. "Normally it is quiet until January 15, then the President submits his budget and makes the state of the union speech."

No word came down from OMB in the fall, so it looked as if AXAF had made it into President Reagan's budget for the 1989 fiscal year. Then on January 11, 1988, NASA Administrator Fletcher was notified that AXAF had been thrown out of the budget, only a few days before President Reagan's state of the union speech.

"It's not unusual for OMB to oppose new starts." Tananbaum said. "It is a closed process there; they don't talk outside of the government. Only to NASA. We were committed. We had support all the way, from Pellerin to Fisk to Fletcher. It is like a game of chicken. People at OMB want to know how important the program is for you. If you take the challenge, they back off."

It's not quite that simple, but then Tananbaum is an incurable optimist. Indeed, as Larry Peterson, who was the visiting senior scientist working in Pellerin's division at the time, put it, "If you're not an optimist, you're dead in that business."

If a program is turned down by OMB, the director of the agency sponsoring that program has the right to reclaim—to make an appeal to the President's budget review board. On January 12, Fletcher made such an appeal to a board that included James Baker, secretary of the treasury, Colin Powell, director of the National Security Agency, Howard Baker, White House chief of staff, and a representative of the Office of Management and Budget. They said no. That left Fletcher with one final appeal—to President Reagan.

According to Fisk, "It was up to Fletcher to carry the water at this point . . . This was not done at my level. We provided lots of information and briefed them and all that good stuff. But the only meeting was the one that he had. I think he probably shared the same sentiment that we needed to get the science program going again and that was how we were going to do it."

Some of the briefing material was pure Pellerin. One briefing board (no transparencies were allowed in presentations to President Reagan) read:

"X-ray astronomy was invented and pioneered in the U.S. We have no X-ray mission now flying, but the Soviets do."

Another said: "The absence of AXAF will give the Soviets and their European colleagues an opportunity to use the Mir space station to make major discoveries in X-ray astronomy."

This concern over lost opportunities was escalated into an implied warning on loss of leadership.

"Space frontiers are being explored. Continued leadership in space science is crucial."

The grand finale dropped any pretense of subtlety. It showed the Soviet hammer and sickle alongside the stars and stripes and a Japanese sun, with the words: "To Whom Will the Future in Space Belong?"

"It was a marketing effort," Pellerin said. "In marketing you try to find what people want. The scientists wanted good science. Fisk wanted good science programs. Fletcher wanted to use the shuttle. What did Reagan want? He wanted to beat the Russians . . . we developed a scenario that the Russians would have the future."

Mercifully, these briefing boards were never shown to the President. The details as to who spoke with whom are unclear—Fisk presumed it was Fletcher to Howard Baker. In any event, on the eve of Fletcher's meeting with President Reagan, word came down from the White House that AXAF was back in the administration's budget.

AXAF had made it through NASA, and it had now made it past the Office of Management and Budget checkpoint with some difficulty. Next came authorization hearings, which were usually not much more than a wave-through. Then came the Checkpoint Charlie of any NASA program: the House Appropriations Committee, whose members were known inside the beltway as the Cardinals of Capitol Hill because of their power and secrecy in allocating federal funds. Jack Kemp once said that "the Appropriations Committee is the most powerful committee in the history of the democratic experience."

In *The Cardinals of Capitol Hill*, Richard Munson describes how the chairmen of the 13 House Appropriations subcommittees and their Senate counterparts control virtually all the discretionary spending of the federal government. In 1988, the Veteran's Administration, the Department of Housing and Urban Development, and the Independent Agencies Subcommittee (VA-HUD panel) controlled the largest amount of discretionary funding for domestic programs, including NASA's budget. The chairman of the VA-HUD panel was Representative Edward Boland of Massachusetts. His chief of staff, or clerk, as he was called, was Richard Malow.

Dick Malow, a graduate of the University of Michigan with a degree in political science, had been a staff member of the VA-HUD panel for 19 years and clerk for 11 years. Over that time he had acquired a formidable reputation as one of the nation's leading experts on engineering and science projects. According to a lobbyist interviewed by Munson, Malow is "a quick study who isn't snowed by the technical jargon of researchers." Another described him as "one of the most significant, adroit, quick staff members on Capitol Hill." A NASA executive said that "Malow has put his mark on almost every project we have. His fingerprints are everywhere."

NASA's high-level managers were in close contact with Malow to get a sense of which projects could make it through the Appropriations Committee and how ones that wouldn't make it could be improved. So great was his influence, perceived or real, that a 1989 *Wall Street Journal* profile of Malow suggested that he was the de facto Administrator of NASA.

Malow scoffs at such hyperbole, but he did have a longer history of dealing with the NASA budget than any of the NASA administrators or other NASA executives.

"I got assigned to NASA almost from day one," he told us. That would have been in late 1972. "I came with a general enthusiasm for NASA. The shuttle was just beginning to be funded. I was influenced by what I saw in the mid-seventies and late seventies as a coming train wreck in the whole budget. I must admit that we put the train wreck off, but it is out there, especially as Social Security comes due."

At first, his belief in an impending budgetary catastrophe inclined Malow to take a hard line on science projects.

"I used to think that we had budget problems, so they [the science projects] could wait."

Then, in the late seventies, he underwent a conversion.

"I convinced Boland to go against Galileo [a $795 million program to send a spacecraft to Jupiter, launched in 1989] and he lost badly in the House."

This rare setback got Malow's attention. He was surprised at the depth of public support for large space science programs.

"After that I became more supportive, a true believer in science, while still taking a hard look at the numbers. I underwent a fairly significant philosophical change . . . After that, my approach was, we have budget problems, so I had to figure out what else we could cut so that we could do the science projects."

In the early eighties, the budget problems only got worse. Congress passed the Gramm-Rudman-Hollings Act, or Gramm-Rudman, as it came to be called. This law required Congress to meet budgetary targets or face automatic spending cuts for all programs. Compounding the difficulty for NASA's science programs was the proposed multibillion-dollar Space Station, which was now eating up a large portion of NASA's budget.

"The bottom line was that discretionary funding for NASA was drastically reduced," Malow said. "I was concerned that the space station would eat everybody's lunch."

Astrophysics programs such as AXAF were hit by a triple whammy—Gramm-Rudman, the Space Station, and the Hubble Space Telescope. The last was especially important when the request to fund a new start for AXAF was put back in the President's budget in January of 1988.

"As we got closer to making a decision on AXAF, it was the background of Hubble that had a great deal of influence," Malow said. "I'll never forget the Hubble request. The first request was for 490 million, then it went up to 700 million, then 2 years later it jumped to 1.2 billion."

When the budget numbers for a program keep climbing like that, the House Appropriations Committee may authorize an investigation.

"We did investigations of various projects, for example tertiary treatment of sewage," Malow explained. "The committee used a separate, wholly independent team—the Surveys and Investigations [S&I] staff—to conduct investigations. Each team normally included a semi-expert or expert and one or two FBI agents. Using the S&I staff, we requested an investigation of Hubble. From that came back a sense that there was a real disconnect between what Marshall [NASA's Marshall Space Flight Center] knew and what headquarters [NASA Headquarters] knew about what was really going on. The problem was that the numbers never held up. NASA's credibility went out the window. That was the worst damage to science. Credibility has a lot to do with it. And that colors what goes after."

The AXAF team scored important points with Malow by being up front about costs. "Right off the bat, you had a feeling that it would cost about the same as Hubble," Malow said. "If they had come in with a $500 million estimate, we would have been suspicious."

Instead they came in with a $1.2 billion price tag, which Malow believed was realistic. Still, he was not fully convinced that another Hubble-class project was in order, given the ongoing budget issues.

Then, Malow relates, "I got a phone call and a visit. This person warned me of a problem with the mirrors."

Who was this person?

"At the time I didn't reveal who it was, and now I don't remember," Malow said. "I got a telephone call from someone else, who mentioned the same problem. They said it was unlikely that it would focus most of the energy within half an arc second, that Einstein had never performed on orbit the way they predicted it would, and that, given the leap in performance required for AXAF, it was unlikely—not impossible, but unlikely—that they could meet the specifications."

The statement about Einstein was incorrect. It had performed fully as well as expected.

Nevertheless, to Malow, "the person who came to me seemed credible. It seemed a little close to the [budgetary] appeal. From my perspective" he went on, "I was always concerned about the budget wall, and I knew that this was a billion-dollar project that OMB had delayed, and then someone comes and tells you that it may not work. What may have put some meat on the bone was that I talked to Fisk. I told him what I heard. I got the sense that testing this mirror might be worthwhile."

Pellerin remembers stronger views.

"Malow said that there was no way that they [the committee members] would support AXAF," Pellerin said. "It would be DOA in the House Appropriations Committee. They had been burned so badly on Hubble and they didn't want to do another one."

Around this time, letters from scientists in support of AXAF began pouring in to Congress.

"Hordes of letters came in," Pellerin remembered. "It must have been a tremendous effort on Harvey's part."

The letters began to have an effect.

"In hearings," Pellerin said, "Len Fisk is subtly pushing for AXAF, in this wonderful way. Congressman Boland said, 'You've got to do something about those goddamn letters.'"

"The letters were very effective," Fisk agreed. "They made Congress want to come to a deal on this and get the new start."

The scientists were pressing Fisk, writing and visiting their Representatives, urging them to give AXAF funding for a new start. Malow was worried about committing to another expensive, possibly overambitious program that might get into trouble. Were they heading for a repeat of the debacle with the Galileo program, wherein Malow's advice had led to an embarrassing defeat for his boss, Congressman Boland, on the floor of the House? It was time to make a deal. Malow scheduled a lunch meeting with Len Fisk, and then another and another. Not surprisingly, given Fisk's reputation for finding the third way, they came up with a unique solution: the mirror challenge.

Simply stated, the mirror challenge was this: prove that you can build the largest set of X-ray mirrors in 3 years, and we will fund the rest of the program. Fail to meet the required specifications within the allotted time and the project will be canceled.

It was, as Fisk would say later, good logic. "Why did Hubble cost so much money?" he asked rhetorically. "It's because the problems came in when there was this huge marching army so people had to solve what were in themselves not expensive problems, but at the same time you had this huge payroll of people. So the strategy was to take the thing that's the most technologically challenging and retire that risk first."

"I liked it," Malow said. "I felt that the challenge would help to focus the project so that you wouldn't get a Hubble that grew without bound. You know the old quote about 'nothing focuses the mind like knowing that you will be hung in twenty-four hours.'"

Before signing on to the deal, Fisk consulted Pellerin and some of the key scientists. He put it to them bluntly. Malow would not support a new start for AXAF in 1988. They could fight it if they wanted to, but if they lost, it would be the end. It would also be the end if they accepted the challenge and failed to meet it.

"Fisk asked me, 'What do you think?'" Pellerin remembered. "I said, 'Oh, my God!' then 'To hell with Dick Malow!'"

Then, as Malow predicted, Pellerin's mind began to focus on the problem.

"I thought, on the one hand, he was not asking us do anything we hadn't said we could do. On the other hand, George Newton was concerned about the progress on the TMA [Technology Mirror Assembly]. We didn't have a clean thing here."

Fisk said to Pellerin, "You said we could do it, and Dick Malow will keep his word. I don't want to go back into the queue."

Tananbaum's first instinct was also to take the fight for AXAF to Congress. "We had support in the Senate. We debated as to whether we should accept the challenge. Fisk urged that we accept it. He was very effective in building a consensus among the scientists."

In May of 1988, Fisk told Malow that they had a deal. The AXAF program would get enough money to build the largest set of mirrors. On or before October 1, 1991, the mirrors would be tested. If the test demonstrated that the mirrors could focus most of the energy from an incoming beam of X-rays onto a circular spot whose diameter was slightly less than the width of a human hair, corresponding to an angular diameter of half an arc second, the rest of the program would be funded. If not, AXAF would be history.

15

The Challenge Is Met

IN AUGUST OF 1988, NASA selected TRW, the same aerospace company that had built the Einstein X-ray Observatory and the Compton Gamma Ray Observatory, as the prime contractor for AXAF. Under the direction of Daniel Goldin, the vice president in charge of the Space and Electronics Division of TRW, the company would manage the subcontracts to grind and polish the two largest mirrors, called P1 (for paraboloid) and H1 (for hyperboloid), to align them, and to test them. The grinding and polishing would be done at Perkin-Elmer Corporation (now Raytheon Optical Systems) in Danbury, Connecticut. The mirrors would then be aligned at Eastman Kodak Corporation in Rochester, New York, and shipped to the Marshall Space Flight Center, where the critical test would be performed.

In 1988, Perkin-Elmer enjoyed, justifiably, a reputation as one of the best, if not the best, mirror-makers in the world. Its latest triumph had been the grinding and polishing of the mirrors for the Hubble Space Telescope, said to be the finest mirrors of that size ever made. A different group at the same company had also made the X-ray mirrors for the Einstein Observatory. The mirror blanks, cylinders of a special low-expansion type of glass called Zerodur, were shipped to Danbury from Schott Glasswerke in Germany.

In principle, the grinding and polishing of the X-ray mirrors is a straightforward process. The glass cylinders are first cut to the appropriate length. Next, the outsides of the cylinders are ground and polished so that they will fit into the High-Resolution Mirror Assembly, the apparatus used to hold the completed assembly of mirror pairs in place. Then, the mirror surfaces are ground to the right shape—or figure, to use the mirror-makers' term—needed to bring the X-rays to a sharp focus. Finally the surfaces are polished to the greatest possible smoothness, also necessary for the best possible focus.

In practice, the grinding and polishing of the X-ray mirrors is a technological task of Herculean proportions, as Dick Malow, the political scientist and clerk of the VA-HUD House Appropriations Subcommittee, correctly surmised.

The glass cylinders are fragile, flexible, and heavy, weighing about 500 pounds each. They must be handled with heavy equipment and with extreme care, and they inevitably sag under the effects of gravity. If the mirrors are not put in some sort of support structure, the gravitational sag will introduce unacceptably large errors. If the support structure is not balanced within a fraction of a percent on all sides, the errors will be too large. In a word, the mirrors are a nightmare to handle, to grind and polish, and to measure.

"The tough part of doing mirrors is to measure what you have done and figure out where you need to go," Harvey Tananbaum explained. "You need machines to measure the roundness, machines for the slope, and machines for the smoothness."

These machines could not be acquired from any local optometrist or telescope-maker, or even in-house at Perkin-Elmer, whose business it was to make the best optical systems money could buy. Nothing in existence was up to the task, which was precisely Dick Malow's point.

Art Napolitano, a highly regarded manager at Perkin-Elmer with a reputation for getting a lot out of his team when the stakes were high and the pressure intense, was reassigned from a space station contract to AXAF.

"I went on vacation," he recalled. "When I came back I was told that I was in charge of AXAF. So I went to the optical facility to see what we had in the way of equipment. It was empty!"

Ira Schmidt was given the job of putting everything together. A team was assigned for each task: the cutting, the grinding, the polishing, and four different kinds of measurement. A major problem was the tight schedule.

"We needed to subcontract out a lot of the work," Schmidt recalled. "But a lot of the subcontractors who had the kind of experience we needed simply could not meet the schedule. I spent a lot of time on the road, visiting subcontractors everywhere, telling them about the program. I didn't hesitate to drop names like NASA or to invoke the national interest. In the end, that's what got most of them on board, the national interest."

In all, there were about 40 major subcontractors, and many small businesses involved in supplying equipment for the grinding, polishing, and measuring of the mirrors.

"We exceeded the federal guidelines for small, minority, and disadvantaged businesses every year," Napolitano said, with obvious pride.

While they were pushing to get the facility up and running, Perkin-Elmer decided to get out of the large optics business and sold the Danbury facility to Hughes Aircraft, one of its largest competitors. This caused some "commotion," as Napolitano put it, but that was minor commotion compared to what would follow. In April of 1990 the Hubble Space Telescope was launched and deployed from the Space Shuttle. A few weeks later it was discovered that the Hubble's mirror had been polished to the wrong shape. Although it was off by only 2 microns, or about 1/20th the width of a human hair, the performance of the telescope was severely compromised. The morale at Hughes Danbury Optical Systems, the new name of the company, plummeted. Although different groups of people were in charge of fabricating the Hubble and AXAF mirrors, some of the people that had worked on the Hubble were working on AXAF and they were called to testify in the investigation that followed the discovery of Hubble's flaw.

"We were concerned that the whole team was losing focus," Napolitano recalled. "Their attention was wandering because of all that was going on with Hubble."

The Hughes Danbury team was also concerned that momentum might build in NASA or Congress to cancel or postpone AXAF until the team understood how and why the Hubble problem had occurred. Napolitano and John Rich, the president of Hughes Danbury, decided to be proactive. They asked for an independent blue-ribbon panel to review the methods and procedures that the AXAF team was using.

"It was embarrassing when it was realized what was wrong [with the Hubble mirror]," Len Fisk recalled. "It should have been so obvious that it just sent shivers up and down your spine that this was the same organization. Now in reality, it wasn't the same organization. The company was different. But more than that the people were different."

So Fisk asked the Allen Board, the review panel headed by retired Air Force General Lew Allen, to look at AXAF.

"We asked [the board] to determine whether or not any of the things that they saw wrong with Hubble were being repeated on AXAF. The answer was, 'No.'"

Although the panel gave the Hughes Danbury AXAF team a good bill of health, they did identify some minor organizational problems. They reserved most of their criticism for NASA, which the board felt was pushing too hard to stay on schedule to meet the challenge.

The scientists and engineers working on the project agreed.

"You had to take a lot of different actions without sufficient thought" Paul Reid, the senior scientist on the project at Hughes Danbury said. "It was hard to stay in control of the project."

From a manager's point of view, the pressure helped more than it hurt. "The mandate was good for the project," Napolitano said. "Without the goal, we might never had made it. It put everybody on notice; it focused everyone. They wanted the project to go."

Thus motivated, by late summer of 1990 the Hughes Danbury group had machines or stations for cutting, grinding, polishing, and measuring the mirrors set up in the huge hangar-sized facility called the Blue Room. Hughes Danbury scientists and engineers formulated the basic design concept for measuring machines, or metrology stations, as they were called, and subcontracted the detailed design and fabrication to Cranfield Precision Engineering, Ltd. The floors of the Blue Room were covered with sticky blue paper to pick up dirt and dust from shoes, and environmentally controlled tents covered the individual stations. The grinding and polishing machine held the mirrors horizontally. It rotated at a moderate rate, about five times as fast as the second hand on a clock. A lathe-like machine pushed the polishing tool in and out while a technician on duty brushed on the slurry.

During the grinding phases the slurry was changed to finer and finer grits of aluminum oxide, down to less than a thousandth of an inch. Grinding and polishing tools as small as 7 millimeters (a little over 1/4th inch) to 1 meter (39 inches) long were used. The final polish had to bring any irregularities down to the size of a dozen or so atoms, or a ten billionth of an inch. Otherwise X-rays would bounce off the mirrors at random, producing a grainy image much like the optical image from a sand-blasted mirror. After intensive research and trial and error, it was determined that this could be done with a slurry of cerium oxide made from a recipe patented by Walter Silvernail. Sivernail, a retired chemist, developed the recipe in his basement laboratory in West Chicago.

After several days in the grinding and polishing station, the mirrors were moved through a series of stations designed to measure the circularity of the barrels, then the finely tapered shape of the barrels, and finally the smoothness of the surfaces. The backbone of all these stations was the Precision Metrology Mount. As described by Larry Cernoch and his colleagues George Cheney, Carolyn Vasisko, and Peter Vo, this mechanical masterwork of heavy steel plates and Invar supports, precision steel and sapphire bearings, bubble levels and load cells, supported the quarter-ton glass cylinders without any distortions or misalignments larger than a few millionths of an inch. By comparison, the width of a human hair is about 1,000 times bigger. The whole thing floated on a cushion of air above a 10-by-12-foot slab of granite that was 2 and 1/2 half feet thick, and weighed 38 tons. The granite slab was supported by 6 legs, each about as thick as a telephone pole. The purpose of the granite slab was to provide inertia to prevent disturbances in the building from shaking the mount during measurements.

A typical measurement of the circularity of the mirrors began with the lowering of a mirror with the wide end down through overhead doors into the station, where it was supported on the Precision Metrology Mount, centered, and vertically aligned with gravity to within the width of a human hair. Laser interferometers—devices that use specially constructed mirrors to send light waves on different paths and then recombine them to precisely measure the lengths of the paths traveled by the waves—measured the diameter of the mirror at 3,600 locations as the mirror was slowly turned on a rotary table. The mirror was then raised and lowered into place once again, the measurements were repeated, and then the mirror was raised, lowered, and measured a third time. This process was repeated along the length of the barrel at 10 different places. The mirror was then pulled out, flipped over, placed into the measuring station with the nar-

One of the AXAF mirrors being lowered into the Precision Metrology Station at Hughes Danbury Optical Systems (now Raytheon Optical Systems). (Raytheon Optical Systems.)

row end down, and the entire process was repeated. The procedures for measuring the shape of the barrels and the smoothness of the surface were similarly painstaking.

"By December 1990, 5 months after the Hubble problems, things were going so well that we were predicting that we were going to make our delivery date," Napolitano said. "Then Mr. Murphy woke up. The cross checks [in the circularity measuring station] were not working."

"We couldn't understand it," Paul Reid recalled. "We knew we had to stop work, no matter what the schedule, and analyze the problem."

"There was a big vibration problem," Napolitano said. "It was devastating. We had to figure out what the problem was. We were a lot less parochial than other groups here had been in the past. We asked everyone for help."

As the problem persisted, concern bordering on panic began to set in.

"The sense was, we couldn't lose any more time," Tananbaum said. "Early in January it was clear that we couldn't get there. Lester Cohen [the mechanical engineering wizard that the Harvard-Smithsonian team turned to when problems such as these arose] wanted to stop, analyze the situation, and fix the equipment if necessary. I told them to fix it, and not to worry about Congress and meeting the deadlines, that I would take care of that. It had a calming effect," Tananbaum remembered with a chuckle, "even though it might not have been true. In any case we didn't have any choice."

Intense telephone conferences and meetings with team members at the Marshall Space Flight Center, Hughes Danbury, TRW, Harvard-Smithsonian, and NASA Headquarters ensued.

"The relationship with Len Fisk and Art Fuchs, the AXAF manager at Headquarters, was important here," Tananbaum said. "I called Fisk and told him we would need 2 months to check the equipment."

Fisk consulted Malow. "Len and I agreed that it would not be fair to hold it against them if they failed to meet the schedule because of the measuring facility," Malow recalled.

Of course, as Fisk pointed out, it did matter a great deal. "Succeeding in November wasn't much help because we wanted to be in cycle with the budget," he said.

The AXAF team was given 2 months to fix the facility.

Teams led by Cohen from Harvard-Smithsonian and John Patterson from Hughes Danbury converged on the Blue Room to analyze the problems. They found three seemingly small but significant problems. Electrical cables were binding the rotary table on the circularity metrology station; the reference bar used for measuring the shape of the barrels had slipped by a few thousandths of an inch; and there was the vibration problem. The last was finally traced to a large fan at the end of the building, 100 feet away. The fan was setting up vibrations at a frequency of 18 cycles per second. This frequency just happened to resonate with the fundamental vibration frequency of the Precision Metrology Station, causing it to hum like a large tuning fork. Well, not quite. The vibration caused the optics to move up and down a distance of a few atoms, a few dozen billionths of an inch. Not much, but enough to spoil the measurements.

Isolation dampers, devices that would detect the vibrations and damp them out—think of holding your hand on a tuning fork to stop its vibration—were installed and other adjustments were made.

By April 1, the metrology stations were up and working again, but the pressure was still on.

"We had 2 months to go and 6 months of work to do," Napolitano said. "I called TRW and told them we were going to shut down the project for one day for a retreat to redo the program."

At the retreat the members of the team decided on a new strategy. They would work around the clock, 7 days a week with four overlapping 12-hour shifts. They would step up the computerized process control methodology developed with the test mirrors. Following this technique, they kept track of every conceivable variable in the grinding and polishing process so that they could quickly identify and eliminate significant sources of error. The result was that they had to make fewer visits to the time-consuming metrology stations.

Matt Magida, described as the floor manager for the mirror fabrication at Hughes Danbury, described the effects of this crisis on the Hughes Danbury team.

"A group of about a dozen people came to view it as their mission to make this project succeed," he said. "Our customers [meaning NASA and the scientists at Harvard-Smithsonian] supported us. We had telecons at 3 and 4 A.M. Leon [van Speybroeck] and Lester [Cohen] and Marty

Grinding the surface of the P1 mirror on the automated cylindrical polisher/grinder at Hughes Danbury Optical Systems (now Raytheon Optical Systems). (Raytheon Optical Systems.)

[Weisskopf] would be waked at their homes to discuss problems, and they never objected."

"Tom [Gordon], who was responsible for the operation of the circularity metrology station, and Joe Cerino, and Ben Catching worked almost continuously—60 to 70 hours at a stretch—to get the metrology stations up and running again," Paul Reid said. "Tom would sleep under his desk. Once Ben had to go out to the mall to buy underwear because they hadn't been home in so long."

"It was a battle," Gordon acknowledged. "One stint was 42 hours without sleeping."

"By April, I felt good that we could pass the test," Paul Reid said. "Even though Lester said that he didn't think we would make it. Actually," he admitted, "we didn't really feel safe until June."

"An important key was the software we developed, which allowed process control," Magida said. "Our prediction capability improved dramatically. This allowed us to stay on a station, grinding and polishing, for 30 days."

"Matt Magida was a stickler for process control," Ira Schmidt observed, "and it really paid off."

In the end, though, what really made things happen was the intangible that the computer could never have predicted, that Magida noticed, but couldn't quite explain—the dedication of the people involved.

"They were tremendous," Napolitano recalled. "I asked TRW if we could shut down for Easter and Memorial Day. It would mean a 36-hour slip, but everyone was pushing hard, and I figured we would get more out of the system if they took some time off. The word went out, and within hours, we got word back from Mahesh Amin [the director of shop personnel] and other people on the floor that they would work without pay, if that was the problem. Mahesh had organized volunteer teams to work the holidays."

With this espirit de corps, Hughes Danbury finished the mirrors 2 weeks ahead of schedule. Reid and van Speybroeck analyzed the final metrology runs and concluded that the mirrors would concentrate at least half of the energy of an incoming beam onto a spot with a diameter of 50 microns, about equal to the width of a human hair. Put another way, the mirrors would have a resolution of half an arc second, seven times better than that of any previous telescope, and equivalent to the ability to read a newspaper headline at a distance of half a mile.

This performance would meet the flight requirements, but van Speybroeck wanted more. Since they had 2 more weeks, why not polish the mirrors more to make them even smoother? The extra polishing posed only a small risk and could significantly improve the utility of the mirrors for scientific investigations. In a move that surprised and delighted the sci-

Quality assurance inspectors performing a visual inspection of the P1 mirror surface at Hughes Danbury Optical Systems (now Raytheon Optical Systems). (Raytheon Optical Systems.)

entists, Fisk agreed to let Hughes Danbury polish the mirrors for an extra 2 weeks.

Why did he take the risk? We posed the question to Fisk 3 years later.

"You try, when you're in that position, to stand back and say, 'Five years from now, what would you like to have done?' You've got to be careful that you don't just take the expedient way out and regret that you did one thing and didn't do something else. We would have all looked kind of foolish if they cracked the mirror, so I'm eternally grateful that the risk was not very high and we got a better mirror out of it. One of the joys of that job is that you do make calls every now and then, and if they're the right ones, you feel smart. If they're the wrong ones, you—it's like being a football coach."

By the time the extra polishing was completed, the mirrors had to be shipped without going to the measuring stations, so none of them knew just how well they had done—or if they had messed something up. The two 500-pound glass mirrors were loaded with care onto an Air-Ride van and shipped to Eastman Kodak in Rochester, New York.

"The truck did dry runs to Rochester," Tananbaum told us. "They didn't want any surprises when they were carrying a $100-million piece of glass."

Before the mirrors arrived in Rochester, the Kodak team had rehearsed the assembly process extensively for 6 weeks using aluminum surrogate mirrors. Actually, they had been preparing for the arrival of the glass for 3 years, designing under John Spina's management an apparatus that would hold the mirrors in alignment, yet allow the relative alignment to be fine-tuned during the test.

"We had to ensure that it would work," explained Gary Matthews, Kodak's chief mechanical engineer on the program, "so we made it so that we could make fast changes in case something came up."

Instead of securing both mirrors, the team put the second mirror, called H1 because it was hyperbola-shaped and would be the outer mirror when the mirror assembly took its final form, on a separate platform with actuators that could move the mirror in steps as small as four millionths of an inch.

Kodak implemented this plan in only 50 days, much more quickly than the original schedule had called for, but after the troubles at Hughes Danbury, the schedule had been compressed to make the September deadline for the test that would make or break the program. The assembled mirror pair was shipped to the X-ray Calibration Facility at the Marshall Space Flight Center in Huntsville, Alabama.

The test facility was constructed to duplicate the conditions under which AXAF would operate–namely, outer space—as closely as possible. A smaller facility had been constructed at Marshall to test the Einstein X-ray Observatory about a dozen years before, but the much more exacting specifications for AXAF required a major upgrade. A huge vacuum chamber, 24 feet in diameter and 75 feet long, was constructed that could be pumped down to less than a billionth of atmospheric pressure.

Normal conversation in the building would produce vibrations a few

Overview of the X-ray Calibration Facility. (NASA/MSFC.)

thousands of an inch in the floor on which the mirrors would be sitting—unacceptably large, so the floor was set on piers on a 5-foot thick concrete slab that was sitting on 2 feet of sand. The X-rays that the mirrors would focus would travel from a carefully controlled tube a third of a mile away. This tube had to be kept at extremely low pressure so that the intervening air would not absorb or scatter the X-rays and thereby compromise the test. The construction of this facility in time for the test was a major engineering achievement by the Marshall engineers under the direction of Danny Johnson.

Once the mirror assembly arrived at the calibration facility, the team from Kodak began setting up for the test. A team of scientists and engineers from Harvard-Smithsonian lead by van Speybroeck, the AXAF telescope scientist, was there, as well as the resident Marshall science and engineering team, which would run the test facility. The major technical difficulty was expected to be the precise alignment of the two mirrors to an accuracy of one ten-thousandth of an inch, about the width of one of those invisible flu viruses that make us so sick every winter. The Kodak group felt ready.

"Using the actuators," Gary Matthews explained, "we would be able to tilt the mirror in the vacuum tube according to the prescription given to us by SAO [Smithsonian Astrophysical Observatory, the Smithsonian part of the Harvard-Smithsonian Center for Astrophysics]. SAO would tell us where to put it, and we would move it there to get it into focus."

Then he laughed, "Of course that didn't work."

The tests depended on complex computer codes called ray-tracing codes. Van Speybroeck was a pioneer in the development of these codes, which sent hypothetical beams of X-rays reflecting off theoretical mirrors to produce a theoretical image. The theoretical mirrors were modeled to represent the actual mirror as nearly as possible. The code included variables to account for the surface roughness of the mirrors, their shapes, and the distortions in shape produced by changes in temperature, the effects of gravity, and the effects of the support structure. It also took into account the spreading of the incoming X-ray beam in the test facility and the effects of any residual air in the vacuum tube where the test was conducted.

At the beginning of an experiment, the air temperature and pressure in the test facility and the characteristics of the test beam of X-rays were all measured. This information, along with the measurements from Hughes Danbury on the shape and smoothness of the mirror, and the information on the alignment of the mirrors were fed into the ray-tracing code. The code, which could take several hours to run on a powerful computer, predicted the expected shape of the X-ray image that was formed on an electronic detector at the focus of the mirrors . In this case, it was expected to be a very small circle, less than a thousandth of an inch in diameter.

Once everything was in place, the test began over the Labor Day weekend of 1991, when workers pumped down the vacuum chamber to a billionth of atmospheric pressure. This took 8 to 12 hours. Then the experimenters waited another 12 hours for any currents in the small amount of residual air left in the chamber to die out. On a signal a switch was thrown

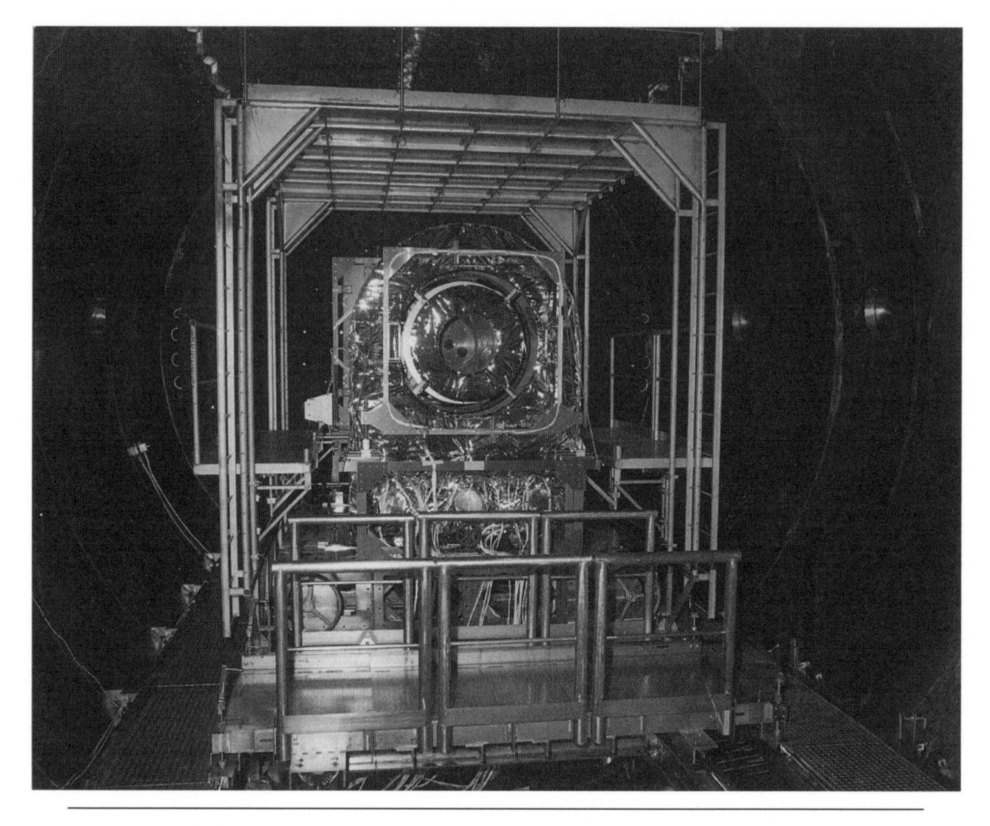

X-ray Calibration Facility vacuum chamber with the mirrors and X-ray detector assembly installed. (NASA/MSFC/A. Byford.)

on the power source, a beam of particles was accelerated to a speed equivalent to that of million-degree gas, and a burst of X-rays began streaming down the tube at 186,000 miles per second. Two millionths of a second later, the mirrors focused the X-rays onto the detector. The signals from the detector were relayed to a bank of computers in the control room, and the ray-tracing experts began to crunch the numbers to see how good the mirrors really were.

Mark Freeman, an engineer with Harvard-Smithsonian, was one of the number crunchers.

"We were doing one-dimensional scans across the detector," he recalled. That is, to save time, they were looking at the data a strip at a time from top to bottom of the detector. If the image was a fine point, then those strips in the vicinity of the image should show a single peak in approximately the same place. They didn't.

"We were getting double peaks on some scans and single peaks on others," said Freeman.

"At first we were annoyed, because we were trying to get a focus, but couldn't," recalled Jack Hughes, who was part of the Harvard-Smithsonian team.

Van Speybroeck, the telescope scientist who had watched over the mirrors' creation like a midwife for half a dozen years, plus another half dozen years of design work, found the results more than annoying.

"Those first results were very, very frightening," he said. "The bad results went on for a few days. We desperately hoped that it was an alignment problem."

The reason for van Speybroeck's alarm was clear—the Hubble Space Telescope. If the problem did not lie in the alignment of the mirrors, then it almost certainly was due to their shape. Had the Hubble experience been repeated? Had the mirrors been polished to an unprecedented smoothness, but the wrong shape? Van Speybroeck, who was still in Cambridge at the time, and Freeman crunched more numbers, varying the tilt of the mirrors one way and then another, hoping this would reproduce the strange results. It didn't.

Dan Schwartz, who was in charge of the Harvard-Smithsonian contingent at the Marshall Space Flight Center, suggested that they take the time to scan the image pixel by pixel. It would be time-consuming and they were short of time, but they had little choice.

"Dan had to leave to go to a meeting in Germany," Hughes remembered. "Ed [Kellogg]) was working on the schedule for the rest of the test. Roger [Brissenden] and I were left in charge, and we were the most junior scientists on the team!"

Word of the problems spread like wildfire. Van Speybroeck traveled from Cambridge to Huntsville to continue his work with Freeman and Bill Podgorski, another Harvard-Smithsonian mechanical engineer.

"We were against the wall," Freeman said. "We had a week or 8 days to figure out what was going on until the congressional deadline."

"At a certain point it was getting very crazy," Hughes said. "There's

Leon, there's me, there's Marty Weisskopf and more people from the project coming down. There wasn't time or space to do our calculations to figure out what the problem was. Roger and I are trying to do the tests and get more observations. I would take the afternoon and evening runs. We would stop at midnight, and I would do my calculations. I remember being there one morning at 5 or 6 A.M., and I got a phone call from Harvey."

Tananbaum asked, "How are things going? Think I should come down?" Hughes responded, "You should absolutely come down." Tananbaum arrived the next day.

"When Harvey got here, things got a lot better," Hughes said, "He found space for us to work and made a plan so we could get our work done, and things moved forward."

"He was a big help," van Speybroeck agreed. "There were too many people wanting to know what the problem was, and you couldn't get away to think. About the only thing they did was to trip over a cord and disable my computer at one point."

When the results from the two-dimensional scans came in a couple of days later, they showed a distorted cross pattern.

"I can remember scratching our heads for a couple of hours," Freeman said. "Jack said that I think we have astigmatism in that spot. That's when we went back to the model."

Astigmatism is a distortion of images caused by incorrectly shaped lenses or mirrors. That was the problem with Hubble's mirrors. Was Hughes panicked by this realization?

"During the scariest time we were too busy to worry," Hughes said. "At first it was so confusing that we couldn't even start thinking that something could be wrong with the optics. You never have time to get really scared . . . You know what it's like. You look at this pattern. You know something is wrong, but it's symmetrical, so you know that it should be amenable to a simple approach."

Hughes began modifying the ray-tracing code to an assumed shape of the mirrors, from circular cylinders to oval-shaped ones. What he was exploring was the possibility that gravity had caused the mirrors to sag out of round ever so slightly into an oval shape. That seemed to explain the astigmatism. In the meanwhile, Dave Zissa and Jim Bilbro of Marshall's ray-tracing team had been analyzing the data and had gotten the peculiar cross pattern.

"At that point I was convinced that the problem was an ovalization of the mirrors caused by gravity," said Hughes.

When in space, the mirrors are in a weightless state, so they don't sag. On Earth, either the experimental set-up must compensate for the sagging with supports or the ray-tracing codes must take the sagging into account and figure out what the image would look like if the sagging were not present. The groups from Kodak, Marshall, and Harvard-Smithsonian all had developed independent ray-tracing codes as a cross check. Before the tests, the Kodak and Harvard-Smithsonian codes had agreed. Bilbro and Zissa had started on theirs later, so they were not ready for the pretest check. Zissa did comment that his preliminary work indicated a problem with the Kodak code, but since the Kodak and Harvard-Smithsonian codes agreed, no one paid much attention.

Now all groups went over all the codes with a fine-tooth comb, looking for errors. Because of Hughes's insight, they knew where to look, and they quickly found a confluence of errors. The Kodak engineers had started out using stiff supports on the structure, so gravity was not an issue, and its effects were not included in their code. As they modified the support structure to make it more adjustable in the test environment, they forgot to modify their codes accordingly. The Harvard-Smithsonian group had included the effects of gravity, but they were unaware that Kodak had changed the support structure. The net result was the same in both cases—the computer codes could not reveal that the cause of the astigmatism was a gravity-induced sagging or ovalization of the mirrors.

"Very quickly after that our guys got a better gravity model," Hughes said. "They got the current design from Kodak and put it into our ray trace. Sure enough the image came out as a dot."

After the errors in the ray-tracing codes had been found and fixed, the obvious next step was to repressurize the vacuum chamber, put additional supports on the mirror to take out the ovalization, and test it again. By now, however, they were hard up against the September 15 deadline.

"We didn't have time to put the springs on, repress, and take measurements before that deadline," Hughes said. They had to go to NASA management, namely Len Fisk, with the computer results. Ironically, the problems with the Hubble mirror helped here. The Hubble team had become expert at taking out the effects of an astigmatic mirror, and NASA officials understood and accepted the procedure. Using software developed for Hubble, the AXAF team showed that the test results implied that the mirrors would focus an X-ray beam to a dot with a diameter about half of the thousandth of an inch that was required by the challenge.

Fisk was ready to go to Congress with this result, and declare the test a success. One might have thought that this decision, which meant that a major victory, a giant leap forward toward the realization of their dream project, would have made the scientists ecstatic. Not so.

"Although pleased, I still wasn't convinced," Hughes said.

"The program was declared a success on the basis of that prediction," van Speybroeck affirmed. "Some people thought that we could stop there. Some of us wanted to prove that it was really that good. Fortunately that opinion prevailed."

The Kodak team was now working flat out to prepare for such a proof. Tom Casey, the lead systems analyst for Kodak's AXAF team, had been hard at it ever since the first confusing images were obtained at the test facility in Huntsville.

"Gary and I were here one entire night . . . we had to get the answer right away," he recalled. "We waited until everybody had gone home and we took every single Macintosh computer and loaded this ray-trace software on there and had a different case running on every one. We were running around between all the different computers. Trying to keep that straight was loads of fun," he laughed. "By the time the next morning rolled around we had identified the problem."

The Kodak analysis showed that if the mirrors were pushed ever so gently to offset the action of gravity, the astigmatism should disappear. After consultation with Lester Cohen as to how to design a system to do this, Matthews flew to Huntsville to pitch his proposal to the assembled project scientists and engineers. He promised to come back in a week with all the hardware necessary to implement the change. After another hectic week of computer analysis and hardware design, Matthews returned to Huntsville with a team of engineers and "a bunch of springs." While a NASA committee debated exactly how hard to push on the ring holding the mirrors, Matthews and his team, who felt they already knew the answer, set to work installing the equipment.

"There was a meeting, and they were arguing about what the force should be," he recalled. "I was sitting in the corner laughing because it was already done. I said, 'Why don't we pump it down and try it?' I had to get back on a plane, and they were still arguing about what the force should be. I didn't really care, because it was pretty much a done deal."

Mark Waldman, the Kodak engineer in charge of the alignment process, was there. The operator started collecting and displaying the image so they could watch the image change while the H1 mirror was being moved. If they had gotten it right this time, then the image should start out as an out-of-alignment double loop that gradually shrank to a point as the H1 mirror was tilted into alignment with the parabola-shaped P1 mirror.

"Over a period of several seconds we watched the loop image shrink quickly," Waldman remembered. "It was incredible!"

"All of a sudden it looked like an image," said Mark Freeman, who had been among the first to start worrying when the first test runs did not look like an acceptable image.

"It was a miracle," said Matthews, who had spent many sleepless nights since the first problematic images. "It really worked well!"

The final measurements showed that the results were even better than expected. The mirrors focused the beam to a resolution of a fifth of an arc second, two and a half times better than required by the challenge.

The senior team members were called to Washington for a party to celebrate. Jack Hughes and Dave Zissa were awarded NASA Public Service medals for their efforts in recognizing and solving the problem, and there were congratulations all around.

Malow told a reporter, Eliot Marshall of *Science* magazine, that the challenge "worked out pretty darn well . . . If my career had to come to an end suddenly, I would like to go out with that one."

Reflecting on the challenge, Fisk said, "I think that the program got off on very much the right foot as a result of having that deal. I think, aside from the fact that it made the program go, that it was a very good thing to have done."

Tananbaum acknowledged that the challenge was valuable in that it forged the teams from industry, NASA, and Harvard-Smithsonian into one group, but he was concerned about the impact of the launch delay on the budget.

"Since we formally started on this program [the mirror test]," he told Marshall, "three years have gone by and the launch has receded between 2 and 1/2 and 3 years; that's very disturbing."

Fred Wojtalik was more explicit about the consequences of the congressional challenge.

"We had to create an unnatural plan," he explained. "We did not have enough money up front, so we put most of it into building glass . . . We got out of step from the way that things are normally done. We had to meet the congressional milestone, so most of our money was going for that . . . In the meantime we were delaying other activities. If you don't get your money in the year you expect it, then the launch date moves out and the overall expense increases."

Later difficulties would prove that Wojtalik's concerns were well founded.

In spite of these misgivings, spirits were high. The teams had, through heroic efforts, met and surpassed a high-stakes challenge. Then, at a celebration party, Dick Malow got up and made a speech in which he, in effect, told them that the party was over.

"We had a big party," Wojtalik recalled. "And Dick Malow made a speech. He said, 'I've got good news and bad news for you. The good news is, you have a program. You've met the requirement; I congratulate you. The bad news is you're not going to have all the money you expected . . .' I don't know if that was prearranged and people thought we weren't going to make it—I can't believe that—but, in any event, we didn't get all of the dollars."

IV

RESTRUCTURING

16

A Bruising Lesson

THE AXAF SCIENTISTS were about to learn a bruising lesson in big league budgetary politics. While they were working valiantly to meet the mirror challenge laid down by Congress as a necessary and sufficient condition for full funding for their project, that deal was in the process of being undermined by events totally beyond their control.

In May of 1991, the VA-HUD House Appropriations Subcommittee, which controls NASA's budget, along with those of the Veteran's Administration, the Department of Housing and Urban Development, and a few other agencies, received word that mounting budget deficits required that it cut its budget by $1.3 billion for the 1992 fiscal year, which would begin in October of 1991. Robert Traxler of Michigan, the chairman of the subcommittee since Boland's retirement from Congress, suggested eliminating NASA's Space Station. This would save $2 billion, thereby freeing up additional dollars to protect science programs such as AXAF.

The Space Station was budgeted for a total cost of $8 billion when it was first proposed in 1983. By 1991, over $5 billion had been spent, nothing had been built, and the estimated cost had ballooned to $40 billion. In the meantime scientific support for the station had evaporated.

Even Vice President Dan Quayle's blue-ribbon panel, chaired by the aerospace executive Norman Augustine, had panned the design. As an expensive project in trouble, the station was an obvious candidate for cutting. Traxler's committee agreed in a 6-3 vote, and sent the recommendation up to the full House Appropriations Committee, which approved the cut.

NASA officials and the aerospace industry were dismayed at this action, as were congressmen whose states were benefiting from Space Station contracts. Despite this, Traxler and his colleagues on the Appropriations Subcommittee had two strong reasons for expecting their action to stand up when it reached the floor of the House of Representatives. First, the House rarely overrode the recommendations of the Appropriations Committee, and second, the White House seemed to be unenthusiastic about the station.

"We had a sense that the White House was looking for an executioner," an associate of Traxler's told the author Richard Munson. "Clearly we were reading tea leaves, and we read them wrong."

This became apparent immediately as the Bush administration encouraged high-level officials at the Office of Management and Budget and NASA to help Congressmen Jim Chapman, a Texas Democrat, and Bill Lowery, a California Republican, to draft an amendment to the budget bill that would restore funding to the Space Station. NASA agreed that the money needed would come from within its agency. The prospects for the Chapman-Lowery Amendment were brightened by the support of the White House, NASA, and the aerospace industry. At the time Space Station work employed about 25,000 workers around the country. Companies in Chapman's and Lowery's home states had major contracts, as did firms in Washington state, the home of House Speaker Tom Foley, and Missouri, the home of Majority Leader Richard Gephardt.

Space scientists and astronomers, though they privately opposed the Space Station, were for the most part reluctant to criticize NASA in public, since it was the primary funding source for most of their projects. A fractious public debate ran the risk of damaging public support for NASA in general and could lead to an overall reduction in funding. Many of them worked behind the scenes in the hopes that if space science funding was cut in the House, the Senate would make up the difference. Others were not so sanguine.

"What this amendment will do is force NASA to eat its own," declared Charles Schumer, then a congressman from New York.

The House restored Space Station funding by passing the Chapman-Lowery amendment, handing Malow's boss, Traxler, a rare defeat. Funding for space science projects was cut accordingly, and despite a belated rearguard letter-writing and lobbying campaign by space scientists, it was not restored in the Senate.

In the wake of the Senate vote, NASA Administrator Richard Truly, a former astronaut who had replaced the science-friendly James Fletcher in 1989 and who had worked hard for the restoration of the station, acknowledged that "much work needs to be done to restore balance to the agency's budget."

Much work indeed. NASA had saved the space station, but the agency had paid dearly for it. NASA's budget that year grew by only 3 percent, as compared with 13 percent the year before and 15 percent the year before that. With the growth of the Space Station unchecked, deep cuts in the major space science programs were inevitable. The upcoming yearly allocation for AXAF was cut from $211 million to $60 million. The projected launch date was moved forward yet another year, to April of 1999.

That was bad news, but Jonathan Grindlay, who was then chairman of the Space Science Working Group, and Kathie Bailey, who was director of the group, warned in a memorandum that even more bad news might be coming.

"One of the more troubling aspects of the Senate bill was its numerous caps on specific projects," Grindlay and Bailey wrote, and went on to quote from the statement by the House and Senate VA-HUD Committee conferees. In their statement, the conferees directed NASA to submit proposed caps for all its activities and to adjust its "expectations and strategic planning to leaner budget allocations in the coming years."

The implications of the proposed caps for AXAF were profound, but the members of the AXAF team seemed not to be very concerned. After all, they had a deal. Pass the test, and Congress will fund the program. They had passed the test, and they fully expected the remainder of the program to be funded, as promised.

"We knew about the proposed caps," Kathy Lestition, a program analyst at the Harvard-Smithsonian Observatory who worked closely with Harvey Tananbaum, recalled, "but it wasn't clear if or how they would apply to us. It is never as simple as saying we have this much money or we don't have it. There was the question of how the OMB accounts for money authorized but not spent, whether or not money could be shifted from one program to another, or from one agency to another. And, it was very difficult to get a straight answer."

At Len Fisk's level, the answer was clear, disturbing, and paradoxical. Under his leadership, the Office of Space Sciences had been too successful on the one hand and not nearly successful enough on the other.

"We had three banner years in '89, '90, and '91 in selling the programs," Fisk told us. "We sold AXAF, we sold CRAF/Cassini, and then we went for the whole enchilada, and sold EOS [Earth Observing Systems], which has a $17-billion price tag on it. The budget was literally wonderful at that point . . . We really had a phenomenal program, but it was built on a certain set of economic assumptions."

These assumptions would prove to be devastatingly wrong as the state of the economy worsened and deficits mounted. Nevertheless, Charlie Pellerin gave Fisk credit for crafting a bold and visionary plan that would have transformed NASA if it had succeeded.

"I was always amazed at the difference of perspective between scientists and administrators," Pellerin said. "The scientists think NASA should be a scientific agency. Len Fisk was the only person who had the absolute brilliance to hatch a plan to make that occur."

In our conversations, Fisk did not claim to be the architect of any plan other than to get as many good science programs going as possible, within the constraints of NASA's budget.

"It was never discussed," Pellerin emphasized. "Len never told me this. I never asked him about it, but it was so clearly in opposition to what the classic NASA position was or to what the Administrator we all worked for wanted, that even when I saw it emerging, I didn't want to talk about it."

The key to this strategy, Pellerin believed, was the Mission to Planet Earth program, which later became known as Earth Observing Systems, or EOS.

"Len created a situation for a choice to be made. It was probably the only time that there was a real choice to be made and a real probability of success. He created a program called Mission to Planet Earth that had the biggest constituency ever from Earth and space scientists, and was so big that there was not room for it and the Space Station both in NASA's budget. Len hoped that reason would prevail."

According to Pellerin, reason didn't even come close.

"We weren't even in the ball park," he said. "The scientists were toddlers in the arena of politics. The big industrialist concerns rolled us. It was a complete defeat. Science advisors, university presidents, every intellectual resource we could bring to bear, were nothing. Then what we had was a Space Station that was still there."

As soon as the Space Station battle was lost on the floor of Congress, both Fisk and Pellerin knew the space science budget was a shambles.

"We were in an untenable situation where we couldn't afford what we had sold," Fisk said. "We tried not to discriminate against any particular discipline, to see how much we could save on everything."

"It was a no-brainer," Pellerin said. "In the coming budget year, it was very obvious that there wouldn't be enough money. I became convinced that we [AXAF] would be canceled if we didn't de-scope."

Fisk agreed. "This is out of my hands, Charlie," he told Pellerin. "It's bigger than me. You're gonna lose."

The thankless task of taking that message to the scientists fell to Pellerin. He invited representatives from the science community, TRW, the Marshall Space Flight Center, and AXAF program officials from NASA Headquarters to a meeting in Washington in early January 1992.

The scientists knew that something was up, because Pellerin had asked the group at Marshall to look at possible ways to reduce costs on AXAF. Word of this request traveled quickly to Tananbaum at the Harvard-Smithsonian Observatory, who immediately got on the phone to Pellerin. All he could find out from the usually forthcoming Pellerin was that the group was going to look at the overall health of the AXAF program.

"There was some trepidation as to what the purpose of the meeting was," Tananbaum recalled, "and a lot of discussion back and forth while people speculated as to what would be discussed."

The speculation came to an end on January 6, when Pellerin laid it on the line for the elite group of 15 to 20 people. Tananbaum and Riccardo Giacconi were there, as were Claude Canizares, Martin Weisskopf, and Fred Wojtalik from the Marshall Space Flight Center, along with a large contingent from NASA Headquarters, including Art Fuchs, the AXAF manager at Headquarters. Pellerin discussed the deepening national budget crisis—the recession, large deficits, the savings and loan debacle. He said that the end of the Cold War meant that competition with the Russians as a justification for the NASA budget was no longer a compelling argument.

"Charlie was basically painting a landscape so that what followed would be better accepted by those around the table," Tananbaum said.

What followed was worse than any of them had expected. Pellerin told the group that, despite meeting the congressional challenge, despite receiving high NASA priority, despite being strongly endorsed by the nationally recognized science advisory boards, despite having an outstanding, experienced management, science, and contractor team, and a well-understood technical concept, AXAF was in danger of imminent cancellation for the simplest of reasons: lack of money.

The AXAF program, Pellerin explained, was caught between a rock and a hard place. The rock was the cap of between $250 million and $300 million that NASA had been directed to put on funding for AXAF in its peak spending years, so that NASA could meet its other budgetary obligations. The existing budget called for funding that averaged about $90 million a year over the required caps for the fiscal years 1994, 1995, and 1996. This was unacceptable. The hard place was the total cost of the program, the "run-out" costs. If the program was revised to stay within the required caps, then the launch date would be pushed forward 2 more years and the total cost would go up by about $300 million. This was also unacceptable.

Pellerin proposed two alternatives for the scientists and program managers to consider. The first would split AXAF into two drastically reduced missions; that would cut the overall cost approximately in half. These missions, however, would represent little or no improvement over the Einstein Observatory or other X-ray telescopes that were currently in line to launch ahead of AXAF.

The second alternative was more realistic in that it would represent a significant improvement in capability over past or scheduled missions at a 30 percent reduction in cost. This alternative also called for launching two separate missions, one for high-resolution imaging and one for precision spectroscopy (measuring the energies of X-rays precisely). The imaging mission would have three mirror pairs rather than six, as originally planned, and the observatory would be launched by a Titan rocket into a high Earth orbit. A high Earth orbit would make for more efficient observing because the Earth would be in the way only a small percentage of the time. A serious drawback would be the inability to service the observatory with Space Shuttle flights. This was not considered to be a detriment by the government accountants, however. Servicing costs money—a lot of money.

Pellerin's comments precipitated first dead silence and then an outburst of questions, not all of them friendly.

"Everybody was reeling," Tananbaum said. "We had after all just met this key milestone a few months before and we had a program that was very sound and we were very capable of doing it."

"We felt that the observatory was almost virtual," Weisskopf recalled. "It was the carrot at the end of a stick that was always the same length. The launch was way in front of us and we would never get to it."

Pellerin pressed ahead, asking the group to open up and consider a wide range of possibilities.

"Some people looked at it as a 'what if,' blue sky type of discussion," Tananbaum said. "In retrospect, it was naive to think that this was just a 'what if.' There was a certain amount of possibilities being put into play, and once put into play there was no chance that it was going to stop."

Tananbaum and others challenged the basic premise of Pellerin's argument.

"How did we know that we couldn't get the funding?" Tananbaum wondered aloud. "Would it be premature if the word got out that we were considering de-scoping?"

The worry was that the idea of restructuring the program would become a self-fulfilling prophecy.

"We were afraid that once we let that genie out of the bottle, we couldn't get it back in," Tananbaum said.

The group disbanded without agreement on a plan of action other than to meet again in about 6 weeks with a larger group of scientists. During the intervening weeks the group members would do feasibility studies on various options.

17

Conflict and Compromise

THE SERIOUSNESS of NASA's budget crisis was confirmed by Norine Noonan of the Office of Management and Budget at a February 12, 1992, meeting of the NASA Space Science Advisory committee. She warned the committee that NASA could expect no growth in its budget for the coming year and endorsed the cancellation of a large space science program, the Comet Rendezvous Asteroid Flyby (CRAF) mission.

On the same day, Richard Truly was fired as the Administrator of NASA, a victim of the mess created by the Space Station that he had championed, and a desire by the National Space Council chaired by Vice President Dan Quayle to get NASA moving again with "smarter, quicker, and cheaper" missions.

Two weeks later, Pellerin made a presentation to the science working group for the AXAF program. He once again emphasized the seriousness of the budget crisis and the need for a less expensive version of the observatory. This time, there was no invitation for "blue sky" thinking, but rather a request to endorse a mission, known as AXAF-E, for its use of expendable rockets to put AXAF in a high Earth orbit. In this AXAF-E version the number of mirror pairs had been cut to two, and two of the original science instruments, both designed to make precise measurements of the energies of the X-rays, had been deleted from the payload.

Tananbaum responded with an effective counterattack. He questioned Pellerin's analysis of the political climate, saying that there appeared to be strong support for AXAF both at the Office of Management and Budget and on Capitol Hill. He maintained that AXAF-E would be a pale imitation of the mission as it had been originally conceived and sold to the scientific community and Congress. He then urged support of another alternative, a modified version called AXAF-M′. The modified version called for the deletion of the same two science instruments, but would have the full complement of mirrors and stay in low Earth orbit, where it could be serviced. Such a mission, he argued, would retain the scientific integrity of the original version of AXAF, and with a change in the management approach could be built at a cost not much greater than Pellerin's AXAF-E.

For 2 days, the scientists debated the merits of the two missions. Pellerin, pressing for some kind of endorsement from the group, asked for a vote on whether AXAF-E would or would not be capable of performing good science. Tananbaum, Weisskopf, and others responded that he had not given them enough information about the cost of the E mission for them to evaluate whether it would be completed sooner or at an appreciably smaller cost. Pellerin kept avoiding the issue of cost and schedule and pushed for an answer on the science.

To the surprise and chagrin of most of the scientists, Riccardo Giacconi sided with Pellerin. As the founder of the field of X-ray astronomy, the guiding force behind the highly successful Uhuru satellite and the Einstein X-ray Observatory, and the originator of the original concept of AXAF, Giacconi commanded respect bordering on reverence from the scientists in the room. He was Tananbaum's mentor and had handpicked him as his successor as head of the Harvard-Smithsonian group. Giacconi argued forcefully that time and money were against them. Other competing missions from Europe and Japan were on the drawing boards and would steal AXAF's thunder if the scientists got into a prolonged battle to preserve the original mission. They had to face the reality that the AXAF mission had to be drastically reduced, endorse Pellerin's plan, and get some mission launched before it was too late.

With a legend on his side, one would have thought that it would be easy for as canny a bureaucrat as Pellerin to get the endorsement of the group. And it would have been, 10 years earlier, before Giacconi left Harvard-Smithsonian to become director of the Space Telescope Science Institute. During that time, a majority of the members of the science working group, especially Tananbaum, Weisskopf, Claude Canizares, Leon van Speybroeck, Dan Schwartz, and Steve O'Dell, had been through a lot together. They had pushed for the new start, and toiled long and hard to meet the mirror challenge. They were dedicated to putting a great observatory in orbit, and they would not yield easily.

On the second day, Pellerin asked the scientists once again to vote on whether AXAF-E would perform good science, apart from cost and programmatic issues. A majority of the group indicated that Pellerin's proposed E mission was not scientifically viable.

Pellerin then changed tactics. He characterized the members of the science working group as a bunch of insiders who were out of touch with the outside community of scientists—an interesting point of view from someone who had worked as a NASA manager for more than 20 years.

Weisskopf rejoined by pointing out that Pellerin had asked them to vote only on the scientific merit of the E mission, and they had determined that it was not good enough.

Pellerin changed tactics yet again, agreeing that maybe AXAF-E wasn't good enough, and asked the group to come up with something in between the E mission and the M′ mission that Tananbaum was supporting. He revived the idea of two smaller missions, one with two mirrors suitable for making high-resolution X-ray images, the other with large, low-resolution foil mirrors and an innovative high-resolution X-ray spectrometer. This was called the E1+E2 program. The scientists, mindful that NASA was unable to keep its promises, were skeptical. If the second mission was canceled at a later date, as they expected it would be, then they would be left with Pellerin's E mission, which they had rejected as inadequate.

A straw vote was taken on the various options. Of the 15 scientists present, 2 voted to stick with the original mission and fight it out, 1 voted to go with Pellerin's E mission. Four chose E1+E2, and 8 chose M′. In a second round, the vote was 10 for M′ and 5 for E1+E2.

After more discussion of the relative merits of these two alternatives, the group adjourned.

During the month of March, both Pellerin and Tananbaum worked the science and political community to solidify support for their positions. Pellerin directed the scientists and engineers at the Marshall Space Flight Center, TRW, and an independent contractor to study the cost effectiveness of putting AXAF onto an expendable launch vehicle.

"I think that Charlie expected them to come up with something magical," Tananbaum said, referring to potential cost savings. "There was no magic in the process. If there was, we probably would have found it long ago."

Tananbaum organized a study of other options, with van Speybroeck running computer simulations of mirror performance for every imaginable case. These results showed that the scientists needed a minimum of three mirrors, preferably four, to have an acceptable mission. This version of what was acceptable was unacceptable to Pellerin, because he felt that the cost of making the mirrors was what would drive the cost over the edge.

"This led to a lot of friction between Charlie and me," Tananbaum acknowledged.

Pellerin agreed with that assessment.

"To say that Harvey was opposed [to a drastic re-scoping along the lines of the E mission] would be a gross understatement," he told us with a laugh—3 years later.

"I didn't want to give in because I thought he was cutting too deeply into the science," Tananbaum explained.

For his part, Tananbaum continued to insist that a modest change in the mission, together with significant changes in the management approach, could accomplish the cost cuts and preserve the scientific value of the mission. For example, lower the cost of making the mirrors, and give Hughes Danbury a chance to produce the mirrors at this cost. If the Hughes Danbury team succeeded, the scientists would have a scientifically strong mission. If the team failed, they would be no worse off than with Pellerin's plan.

Tananbaum's point was that by taking the safe route proposed by Pellerin they would be spending 75 percent of the money, but delivering only half the science. To prove this point, he had the Harvard-Smithsonian group estimate the costs for a hypothetical mission in which only one pair of mirrors, the ones already built, were used. They found that this mission would cost only 10 percent less than the originally planned mission with its full complement of six mirrors.

"It was like asking us to settle for a Hyundai when we could get a Mercedes for only a thousand dollars more," van Speybroeck remarked.

The argument won more people over to the side of the scientists opposing Pellerin's proposed mission.

"Even Art Fuchs, my program manager, sided with Harvey," Pellerin exclaimed. "He got ticked off at me. Everyone did. I think what probably happened was that I was out there too far in front of them. I was impatient and they were frustrated. This led to a period of bad relationships. It got to be a very divisive period."

"I think Charlie probably had a view of where things were headed, but it's very hard to tell sometimes how much is true prophecy and how much is self-fulfilling prophecy," Tananbaum observed.

Pellerin kept pushing his position. When a report reached Tananbaum that Pellerin had told an influential group of scientists that AXAF was dead, Tananbaum asked Pellerin to explain.

"It turned out that he meant that the original version of AXAF was unsupportable," Tananbaum said. "This bothered me . . . saying that it's dead to somebody that's not up to speed . . . could be interpreted as meaning it is *dead*," he said. "It doesn't need Congress to kill it, or the agency to kill it."

Tananbaum became even more determined.

"We had everything to lose," he continued. "I was perfectly willing to wade in and fight until I was bloody and battered and broken. I didn't feel too bloody, battered, or broken—I was exhausted, I was frustrated, I was angry, I was scared. But we were going to fight this thing out as best we could."

The struggle to retain as much of the original AXAF mission as possible was made more difficult by Giacconi's continued support of Pellerin's position.

"It was probably the only time in a 10- or 20-year period that Charlie and Riccardo were on the same side of the fence," Tananbaum said, shaking his head in bemusement. "Riccardo was influenced a lot by his experience with Hubble, and wanted to get away from servicing and the shuttle . . . he also had better vision than all of us put together, so he was often able to see the end of the game."

Giacconi's vision told him that the scientists would lose this battle, so they should take what they could get, move up the launch date, and get on with it. "I maintain that an E1 program which actually achieves launch within this decade is infinitely better than a virtual AXAF-M′ program," he wrote in a minority report that accompanied Weisskopf's majority report supporting the M′. In the meantime the TRW study came back with an important conclusion.

"A major objective of our study was to reduce the overall weight of the observatory," said Ralph Schilling, who was the lead scientist in the TRW study. The members of the TRW team found that if they used graphite extensively in the spacecraft rather than metal, and epoxy rather than bolts, they could make a strong lightweight spacecraft. This was a major breakthrough. "It meant that we could add mirrors," Schilling explained. "And we could boost the spacecraft to a higher orbit."

How many more mirrors? The total could increase to as many as four. How high an orbit? An elliptical orbit in which the altitude of the spacecraft varied from 6,000 miles to as high as 80,000 miles seemed possible. Such an orbit would make for longer observing times out of the shadow of the Earth and away from the radiation belts that surrounded the Earth. The longer observing times translated into improved efficiency of the observatory. Suddenly a compromise seemed possible. But Pellerin refused to budge from his position of only two mirrors.

"At this point, I got so frustrated with Charlie that I told him that I wanted to work with Len Fisk directly," Tananbaum related. "In fact, I didn't tell him that. I told him I was going to work with other people and I assumed he would figure out that I meant Len Fisk."

Pellerin had walked into Art Fuchs's office when Fuchs was talking to Tananbaum. Fuchs turned on the speakerphone and the conversation became a three-way discussion. Pellerin asked Tananbaum whom he was going to work with.

"In a fairly childish way, I refused to tell him," Tananbaum said. "And in a fairly childish way he stormed out of the office."

Tananbaum and Canizares met with Fisk in early April. A couple of weeks earlier, Dan Goldin, a former TRW executive, had been sworn in as the new administrator for NASA. Because of his previous ties with TRW, the prime contractor for AXAF, he recused himself from any discussion of the restructuring of the project. That left the final decision in Len Fisk's hands.

Fisk quickly cut through the fog of rancor and suspicion that had enveloped the program. He told Tananbaum and Canizares that the crisis was real, that AXAF needed to stay within the yearly caps, and that a mission that could be serviced was out of the question. It was too expensive. The modified, or M′, option for AXAF, the one supported by Tananbaum, Canizares, and most of the AXAF science working group, was not viable.

On the positive side, Fisk assured them that, if they could fit four mirrors into the weight requirements for a high Earth orbit, he would accept that. Further, he would support the option of a second mission devoted to X-ray spectroscopy with the low-resolution mirrors.

Fisk's credibility with Tananbaum and Canizares was such that, in Tananbaum's words, they "crossed over the bridge." The studies of the various options over the previous 2 months played a role in their decision, too.

"Some of the studies had been very good," Tananbaum elaborated. "I think we had been convinced that we weren't going to have servicing. The high Earth orbit, in many ways, was more attractive because the observing efficiency was higher—the Earth didn't get in the way."

Another benefit of an orbit high above Earth is that the spacecraft would spend much less time in the Earth's shadow than it would in a low Earth orbit. This means that the variations in temperature are much less severe, and so the stress on spacecraft components is greatly reduced.

"I think parts of this process grow on you, and as you argue the pros and cons you can see some of the positive aspects of the alternatives," said Tananbaum. "Also the second mission was endorsed by Len as something he would support, so it was no longer something we were putting in as a throwaway."

The next step was to bring the rest of the science working group along.

"This is the best we can do," Tananbaum told them. "Let's make it the best we can make it."

"Harvey's role [in the restructuring] was very important," Schilling said, "because he would act as a catalyst among the scientific community to encourage them to embrace the final solution. He was very instrumental in assessing the scientific viability of the various options that were presented."

In August of 1992, NASA announced that AXAF had been restructured into two complementary missions. One would be primarily an imaging mission, carrying four high-resolution mirrors. It would be launched on the Space Shuttle in September of 1998. The other mission, planned for launch about a year later, would use lower-resolution foil mirrors in tandem with high-resolution spectrometers.

The scientists were united in support of these missions, but divided on the question as to whether the restructuring was really necessary.

Giacconi advocated restructuring from the beginning, primarily because his experience with the space telescope had shown that servicing the observatory in orbit would be exorbitantly expensive. Canizares grudgingly admitted that "Charlie read the tea leaves right on this one."

Leon van Speybroeck was unconvinced.

"Restructuring really gained very little except for dropping of the servicing in exchange for a longer lifetime," he maintained. "It is evidence of a failure of management . . . Charlie Pellerin felt that he couldn't sell the program to his supervisors without convincing Congress that he had made the scientists bleed a bit."

Tananbaum agreed. "Charlie would always say, 'There has to be blood on the floor.' I don't know why it was never his blood. It was always the scientists' blood."

After a moment's reflection, Tananbaum added, "Maybe some of Charlie's blood did get on the floor. Certainly some of Art Fuchs's blood, [him] being between Charlie and me. Art did a tremendous job, keeping us communicating and preventing certain destruction."

Tananbaum, who led the fight against restructuring, probably had more time invested in the program than anyone. Later, as he looked at the restructuring process from every conceivable angle, he was philosophical about it.

"It's possible that Charlie saved the program [by restructuring it]" Tananbaum said. "It's also possible that we wouldn't have had to restructure if he hadn't started the process."

"Maybe there is some sort of sociological law that you have to go through some kind of process like this in order to be convinced that the distilled product is in fact the product. That the changes, even though they are gut-wrenching changes, are changes that you can not only adopt, but eventually embrace and advocate." He paused to laugh ironically. "I like to think that there was some reason that we went through this painful process."

From the point of view of TRW, the prime contractor, it was not that painful a process.

"We welcomed the idea of restructuring the program to provide a stable funding profile," explained TRW's Schilling. "That is what has happened since. We have received the promised funding. Before that, we had come to expect a cut every year, so we would have to replan our work."

As for Pellerin's role, Schilling said, "I think Charlie did exactly the right thing. He did everything he possibly could to encourage the scientists to think broadly about a very wide range of options in terms of capability and cost. If not for his leadership, we might not have gotten there that quickly."

According to Dick Malow, getting there quickly was important.

"There were a lot of questions in Congress about whether it was worthwhile, given the budget issues," he told us. "We could have lost the whole mission. It was de-scoped just in time."

Fisk agreed. "From my perspective," he said, "it saved the program. You'd like to think that people would honor their commitments, but it isn't that way."

"The people arguing against me said, 'We did the mirrors, they will support us,'" Pellerin elaborated. "Scientists do this all the time. They look at something and want to view it as an isolated entity. We knew that the money wasn't there. The power had moved toward the Senate. It's like horse racing. It's not enough to have a horse that you love. You have to have a horse that can win."

Within a year both Pellerin and Fisk had left NASA in the wake of a sweeping reorganization instituted by Dan Goldin. In October of 1993, Congress canceled the X-ray spectroscopy mission. The spectrometer gained a new lease on life when it was chosen to be one of the primary instruments on Japan's Astro-E X-ray Observatory. Unfortunately, it was destroyed when Astro-E failed to reach orbit after launch in February of 2000.

V

BUILDING AN OBSERVATORY

18

Grinding, Polishing, and Coating the Mirrors

By the end of 1992, the grinding and polishing stations at Hughes Danbury Optical Systems were up and running again. The cavernous Blue Room facility had been sitting idle for almost a year while NASA officials and scientists debated the fate of AXAF during the restructuring ordeal. Now, under Ira Schmidt's direction, the grinding, polishing, and testing was going on 24 hours a day, 7 days a week.

Putting to use the lessons learned during the mirror challenge, the Hughes Danbury team refined the computer modeling of the process. They continuously evaluated more than 300 procedures used in the grinding and polishing process and updated them on the basis of suggestions from the operators in the shop or on the floor. Schmidt, Matt Magida, Paul Reid, John Patterson, and Art Napolitano went over these recommendations and the data with a fine-tooth comb in daily planning sessions where the work plan for 3 days ahead was examined and revised if needed. Responsibilities up and down the line for each step were clearly defined, operational problems were quickly identified and documented, and corrective action was taken. Although this close oversight was initially perceived as a threat, the end result was a feeling of involvement and empowerment of the team members as they all worked together to resolve problems. One major bottleneck that was identified by this process was the large number of signatures required by managers and other officers at Hughes Danbury, TRW, and NASA before a procedure could be changed. A streamlined sign-off policy was developed and put into practice, saving everyone valuable time.

The most enlightened and efficient process control in the world will be in vain if the machines break down, so Rita Cebik and Gordon Lester applied the same meticulous procedures to keep the metrology stations running for 500 days without a failure, and Mahesh Amin kept the machine shop running around the clock.

Art Napolitano's office was full of books on a management theory called CMI, for "Continued Measurable Improvement." The articles in the Hughes Danbury magazine *TechNews* invoke CMI like a mantra. It seems like the stuff of a *Dilbert* cartoon, except for one thing: it worked. The performance in polishing and grinding of the AXAF mirrors far exceeded the old standards in the industry. Whereas the previous best examples of mirror polishing had taken 10 cycles of grinding, polishing, and measuring to get to the desired figure and smoothness, the Hughes Danbury team took only 4 cycles on the final pair of mirrors. The team members would start with a mirror that had the desired shape to within a ten thousandth of an inch and make it accurate to within a millionth of an inch. If the mountains of the state of Colorado were to be ground down in the same proportion as the bumps on the AXAF mirrors, Pike's Peak would be less than an inch tall!

"We have learned that optics fabrication is more science than art," Paul Reid wrote in an article describing their remarkable achievement.

After the mirrors were completed and inspected, they were sealed in a bag and put into an inner shipping container built especially for the mirrors by Schott Glasswerke, a German firm. This container was placed inside a larger Schott container. A North American Air-Ride moving van then carried the mirrors to Optical Coating Laboratories, Inc. (OCLI), in Santa Rosa, California. OCLI is recognized worldwide for the precision of its thin-film-coating techniques, used on many devices, from computer screens to the Space Shuttle windows. The drivers, a husband-wife team named Wayne and Jody Radford, made several trips back and forth across the country with the mirrors. The acceleration, pressure, and temperature inside the moving van were continually monitored electronically by a team of TRW and OCLI personnel following closely (but not too closely!) behind in a high-tech motor home. Another vehicle crept along in front of the van, keeping a careful watch for road hazards.

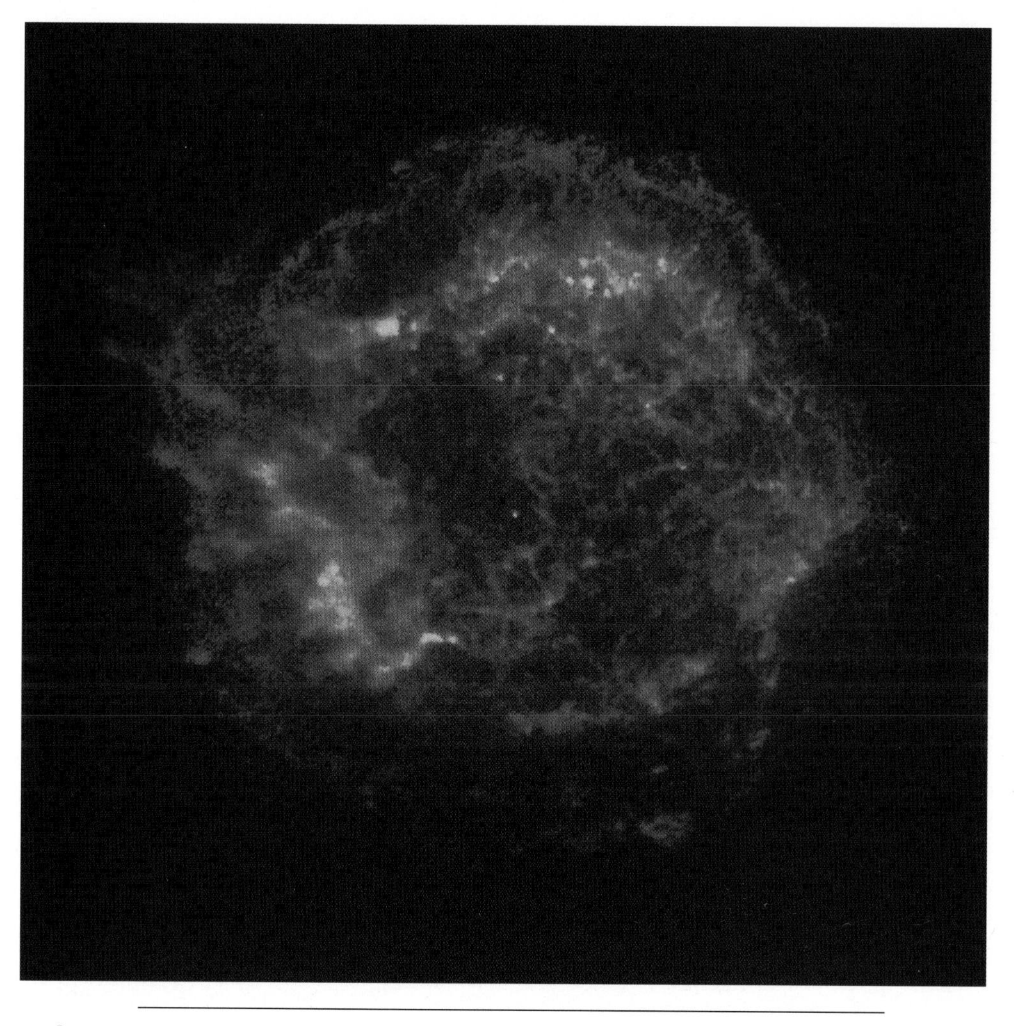

PLATE 1. *Chandra image of Cassiopeia A supernova remnant.* (NASA/SAO/CXC.)

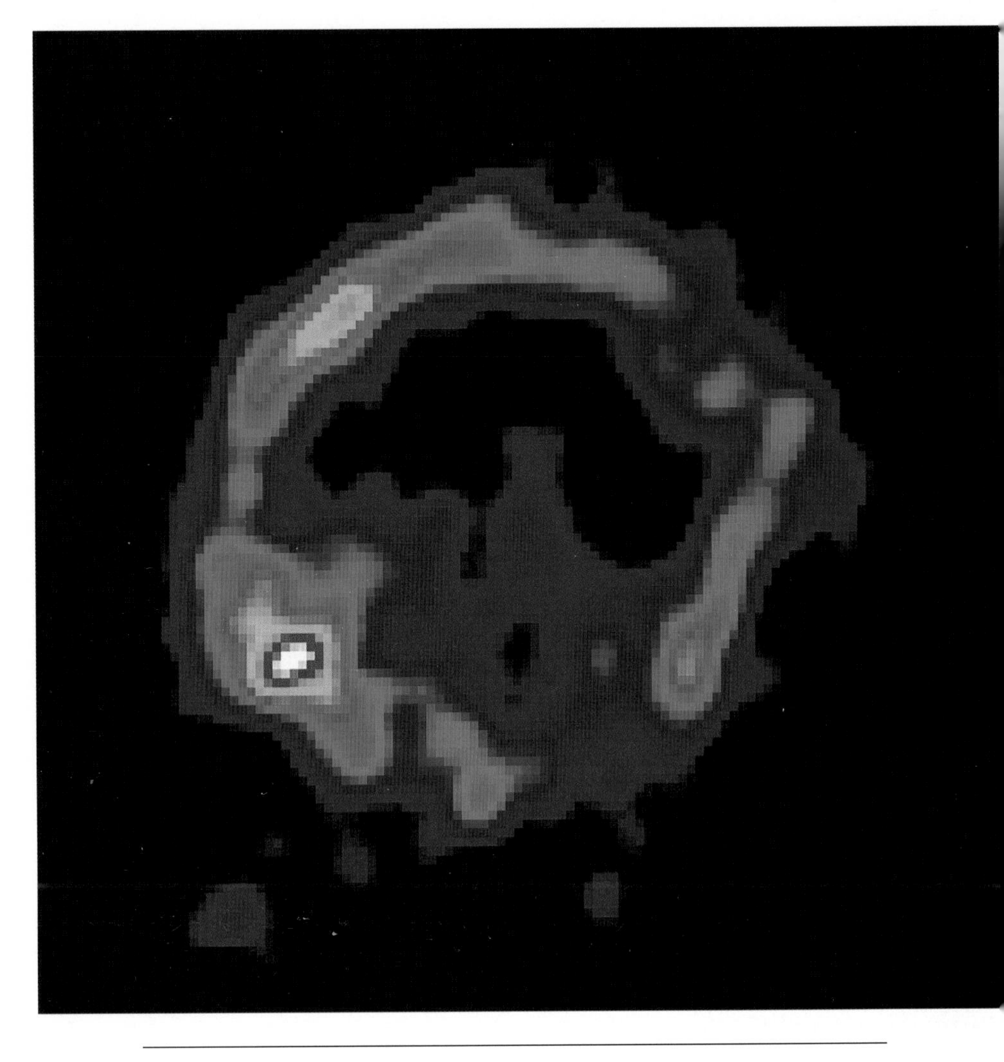

PLATE II. *Chandra image of the supernova remnant E0102–72.* (NASA/MIT.)

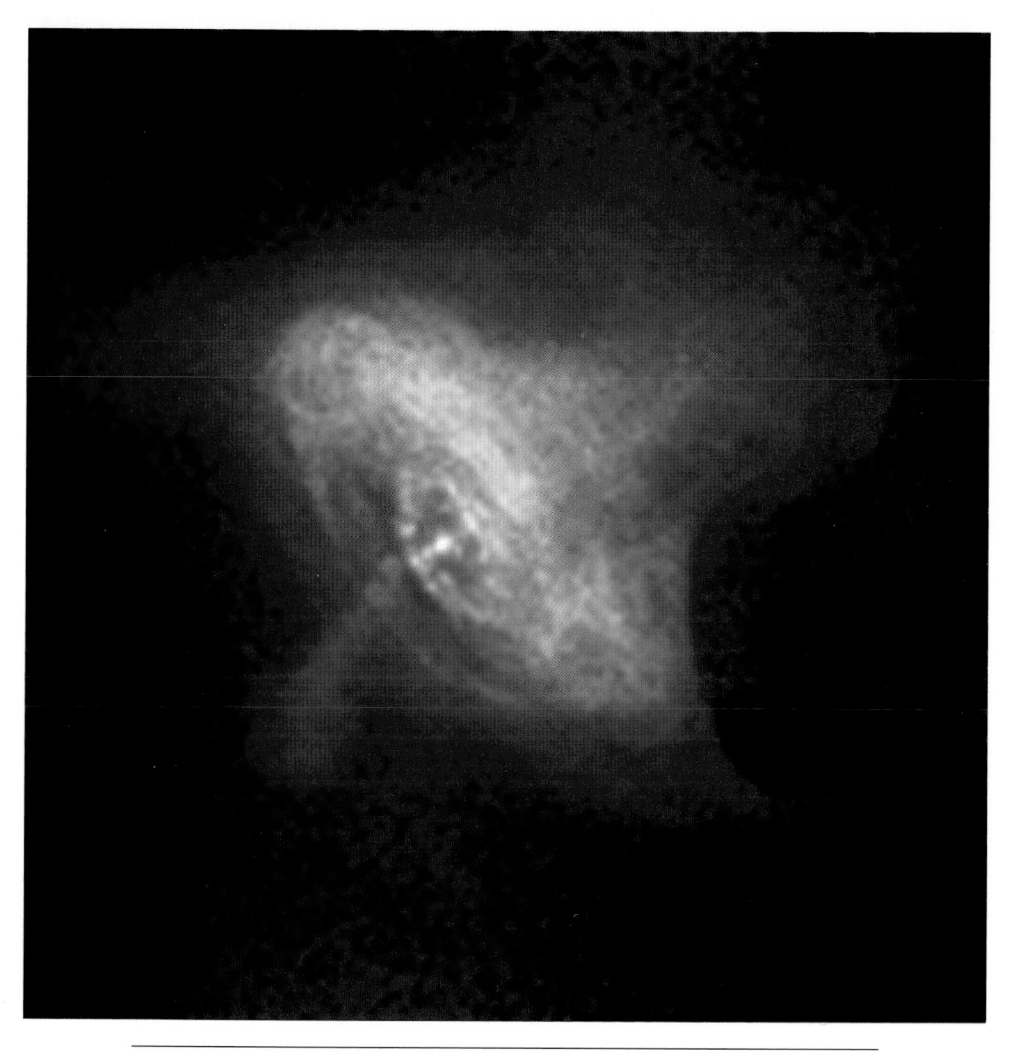

PLATE III. *Chandra image of the Crab Nebula pulsar and supernova remnant.* (NASA/SAO/CXC.)

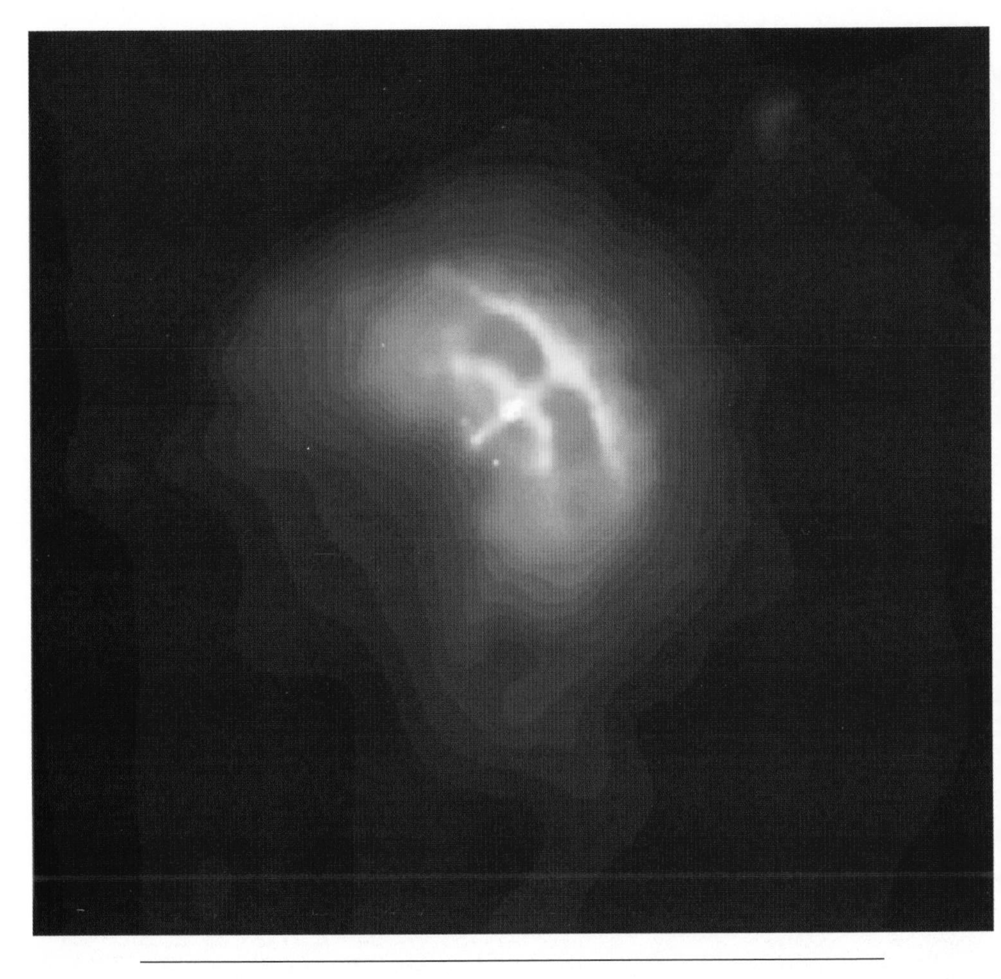

PLATE IV. *Chandra image of the Vela pulsar and nebula.* (NASA/*Pennsylvania State University.*)

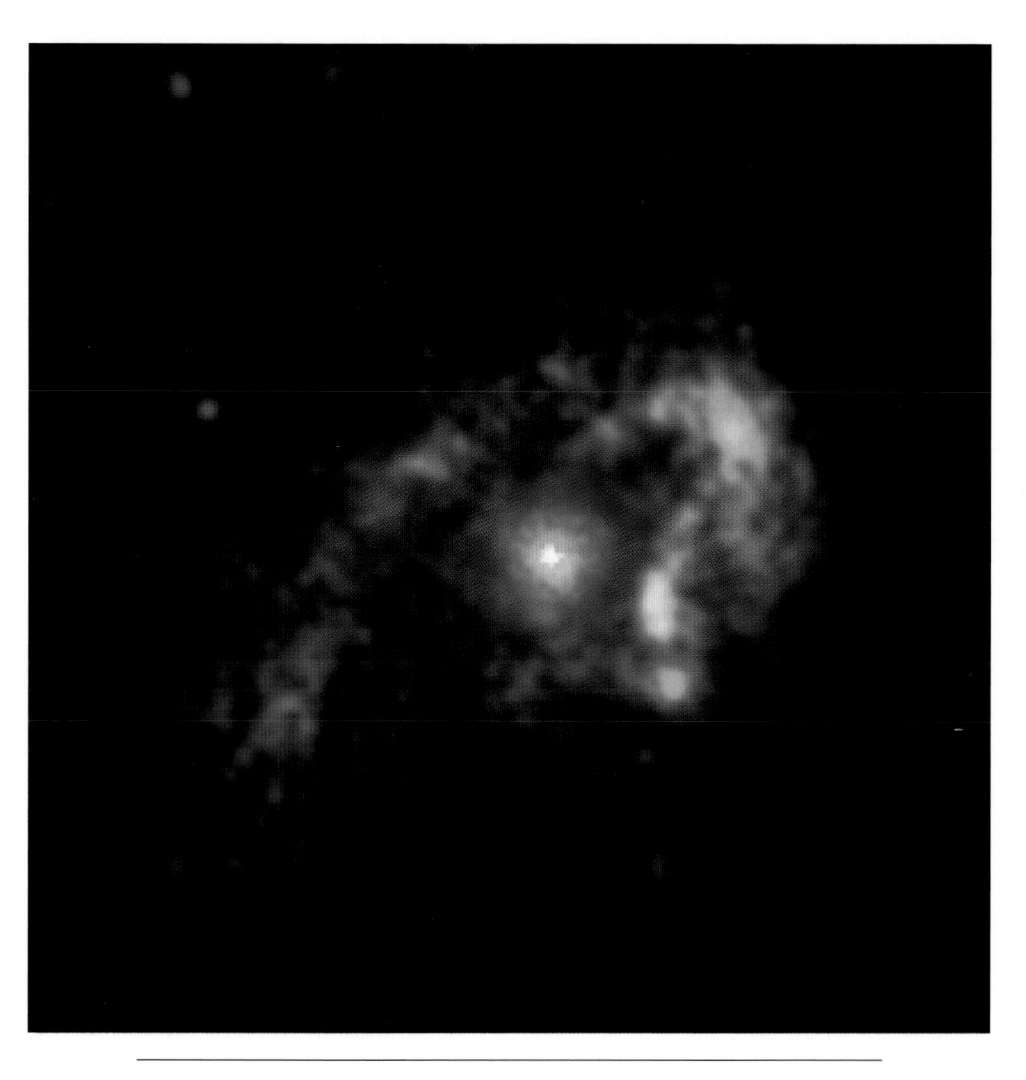

PLATE V. *Chandra image of Eta Carinae.* (NASA/SAO/CXC.)

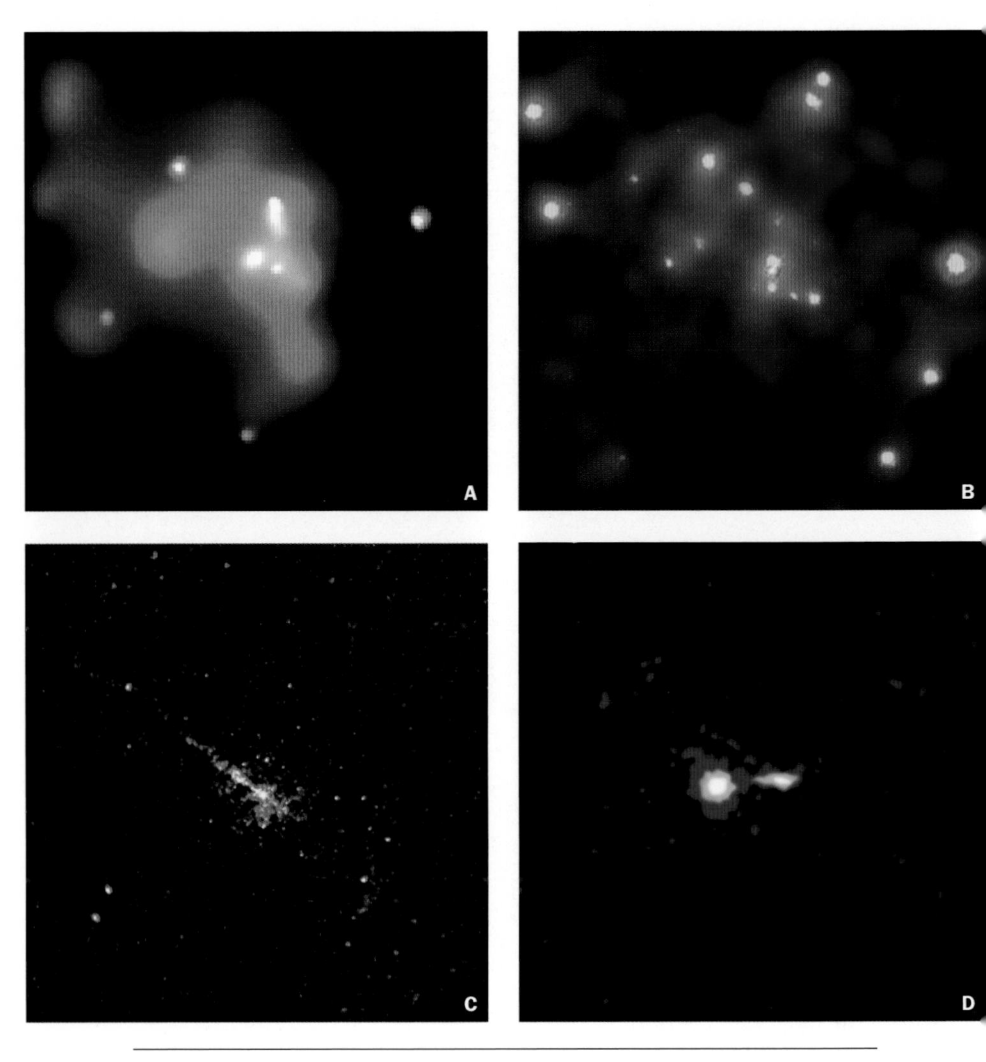

PLATE VIA. *Chandra image of the center of our galaxy.* (NASA/*Pennsylvania State University.*)

PLATE VIB. *Chandra image of the center of the Andromeda galaxy. The blue dot is believed to be an* X-ray *source produced by matter falling into a giant black hole.* (NASA/SAO/CXC.)

PLATE VIC. *Chandra image of the galaxy Centaurus A.* (NASA/SAO/CXC.)

PLATE VID. *Chandra image of the quasar PKS 0637–752.* (NASA/SAO/CXC.)

FACING PAGE

PLATE VIIA. *Chandra image of the galaxy cluster 3C295.* (NASA/SAO/CXC.)

PLATE VIIB. *Chandra image of the galaxy cluster Abell 2142.* (NASA/SAO/CXC.)

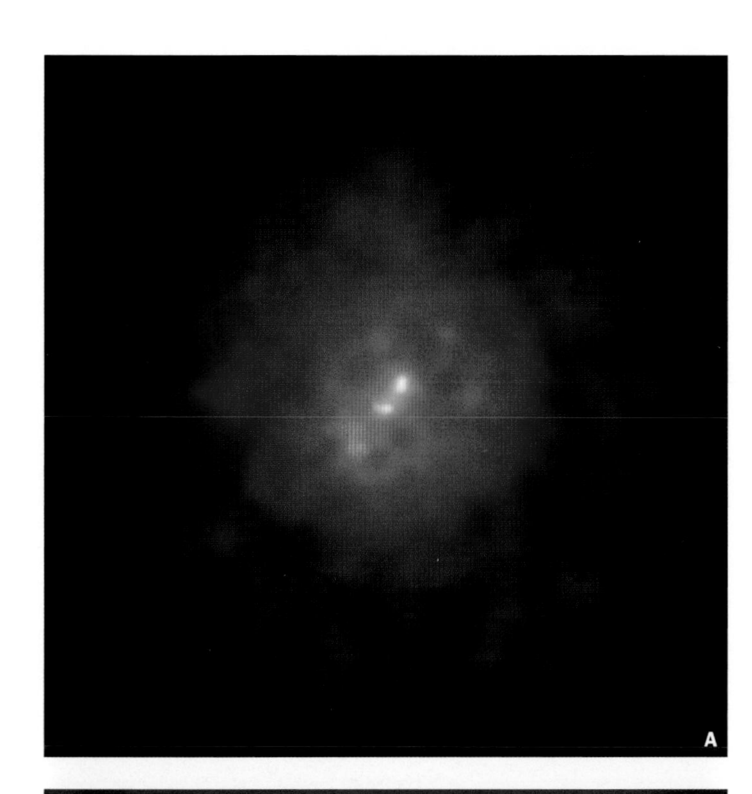
A

B

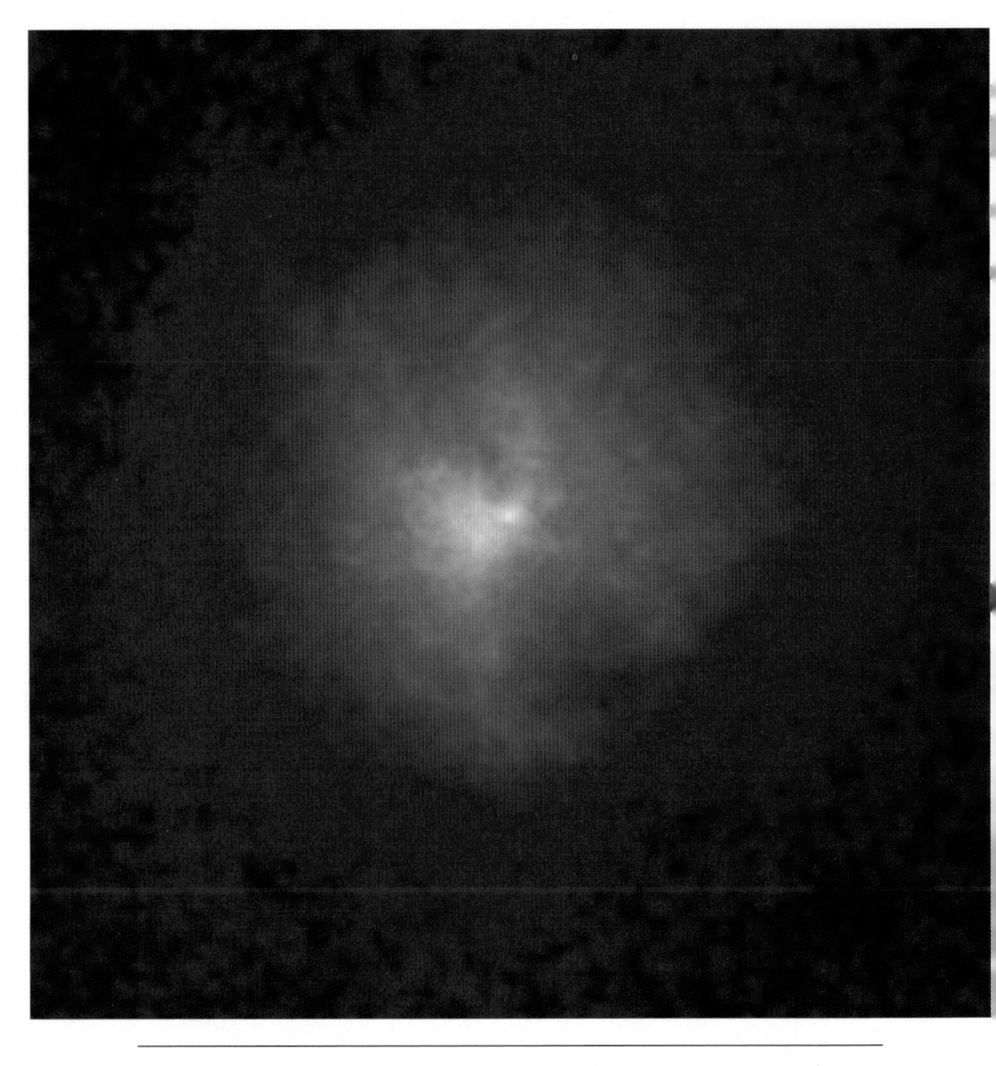

PLATE VIII. *Chandra image of the Hydra A cluster of galaxies.* (NASA/SAO/CXC.)

Loading mirrors onto the Air-Ride moving van. (Optical Coating Laboratories, Inc.)

When a mirror arrived at OCLI, it was treated with all the care that a multimillion-dollar piece of glass deserved. The outer container was removed, and the mirror, still inside one container and the sealed bag, was carefully moved with an electric forklift into a clean anteroom, where the inner shipping container and sealed bags were removed. The mirror was inspected by Andrew Longmire and Alicia Hernandez to make sure nothing had happened to it in transit. Then Bob Jones and Roy Wilson fitted the mirror with a handling fixture designed by Eastman Kodak and moved it into the Class 100 clean room.

In a Class 100 clean room, air is continually circulated through high-efficiency particle air filters until fewer than 100 particles greater than 1 micron in diameter (the size of a cell or a tiny dust grain) flow through a filter 1 square meter in size (about the size of a card table or a 56-inch TV screen) every minute. In contrast, the corresponding number of particles in a clean office would be about 500,000. The cleanest clean room in the mirror fabrication facility at Hughes Danbury was Class 10,000. The clean

Inspecting the mirrors at Optical Coating Laboratories, Inc. (Optical Coating Laboratories, Inc.)

room at OCLI is 5,000 times cleaner than a clean office and 100 times cleaner than the facility at Hughes Danbury. From this point on, anyone in the same room with the mirrors would wear a bunny suit and a gauze mask.

If you have ever painted a wall or a smooth surface, you know that if you don't get it clean first, it will be lumpy no matter how smoothly you apply the paint. In the same way, the extraordinary smoothness achieved by the Hughes Danbury mirror team would have been for naught if the cleaning and coating were not done with equal precision. For X-ray mirrors, lumpy could mean a layer of organic material left by a fingerprint. An X-ray mirror sullied with fingerprints would be equivalent to a normal mirror pitted by a sandblaster. The mirrors had to be 99.999 percent clear of any foreign particles larger than a typical cell.

How do you clean glass to that precision?

"Just like you wash your car," said Bob Hahn, the chief engineer for the AXAF program at OCLI. "You hose it down, wash it with a detergent using a polyvinyl swab, and wash it again. And again. And again."

Once the glass was clean, it required, in the words of Bob Langley, an OCLI engineer, "constant vigilance to keep the mirrors clean." They were moved as quickly as possible into the coating facility.

The coating took place inside a vacuum chamber pumped down to less than a billionth of an atmospheric pressure. The vacuum was necessary because, as Jerry Johnston, the AXAF program manager at OCLI, explained, "We don't want any weather in there. We don't want any turbulence or cloud formation, or condensation onto the glass . . . we have to get it right the first time. You can't remove the coating once it is on."

"Right means smooth to within a few atoms," Hahn reminded us.

Once the high vacuum had been achieved, a thin slab of chromium was lowered into the center of the barrel-shaped glass. A beam of ions—atoms that have a positive charge because an electron has been removed, allowing the energy of the ion beam to be finely tuned—was directed onto the chromium. The ions knocked chromium atoms off the slab. The chromium atoms drifted across the chamber and stuck to the glass. All the while the glass rotated at about twice the rate of the second hand on a clock to ensure that a uniform coating of chrome half a millionth of an inch thick was deposited on the glass. This was the base layer.

The process was then repeated for the reflecting layer. This layer was composed of iridium, a metal more dense than gold. In the course of designing the mirrors for AXAF, the mission support team at Harvard-Smithsonian tested nickel, gold, platinum, and iridium. Mirrors coated with iridium gave the best overall performance for reflecting X-rays. The OCLI team coated the mirrors with a uniform layer of iridium one millionth of an inch thick.

OCLI delivered the last pair of mirrors ahead of schedule to Eastman Kodak in Rochester, New York, in February of 1996. The coating exceeded specifications, making the AXAF mirrors' coating the smoothest ever produced.

19

The Mirror Assembly

"ONE, TWO, three, go!"

On the count, two men and one woman dressed in white bunny suits squeezed the triggers on their guns as an anxious crowd watched through a window.

This was not a surreal replay of the gunfight at the OK Corral, but tack bonding at the Eastman Kodak's clean room assembly tower by some of the steadiest hands in the East. It was a crucial moment in the construction of AXAF. The technicians, Charlie Placito, David Sime, and Yvette Femiano, were dressed in clean room bunny suits to prevent particles or oils of any kind from getting on AXAF's mirrors. Their guns were tacking guns, sophisticated types of caulking guns, that dispensed epoxy into a small port in an Invar pad that held one of the mirrors in its carbon composite support structure.

Previous attempts to secure the pads to the support structure had failed to produce a bond that met the exacting specifications for AXAF. The hydrostatic pressure of the epoxy, though it amounted to less than a fraction of an ounce of force, had caused the mirror alignment to shift by a small, but measurable amount. This was bad news, because, as Kodak's project manager, Jeff Wynn, observed, "What we learned with AXAF was that, if you can measure it, you are probably out of specs."

After that failure, the engineers and scientists worked feverishly with computer modeling to figure out how they could reduce the error budget to an acceptable level. Edward Swigonski, one of the team of crack technicians, had a better idea, one that precision craftsmen– furnituremakers, welders, seamstresses– know well: first tack the structure into place with the smallest possible amount of epoxy, then adjust it to where it is in alignment, and then complete the job. When three people apply the tack bonds simultaneously at equally spaced points around the cylindrical structure, the chances are greatly increased that the original tacks will be nearly right.

The tack bonds, which were about the size of two pin heads side by side, were applied carefully but quickly, and the crew quickly left. During the time they spent in the assembly chamber, their body heat was heating up the air. This small amount of heat rose in the chamber, setting up a circulation pattern much like that created when rising hot air inland generates a sea breeze. These air currents would affect the path of the laser light used to check the alignment of the mirrors.

"Once the crew go in," Charlie Atkinson explained, "their temperature changes the alignment measurement, so we had to wait a couple of hours."

Once the air currents subsided, it would take half an hour to make the measurement, which would tell them whether or not the mirrors had moved in the tacking process, and if so, how much. They then had another hour and a half to decide whether to apply the rest of the epoxy or to pull the mirrors apart and start over again.

"The first time, the mirror moved, and we didn't like it," Gary Matthews recalled. "It was late Friday."

"We had already decided that we wouldn't make the assembly crew work another weekend, that we would abort the assembly and wait until Monday," Atkinson said.

Atkinson and Matthews were sitting in their office, bemoaning the lost schedule time, when the assembly crew came in.

"Well, what are we going to do now?" Placito asked.

"We'll pick it up on Monday," Matthews replied.

"What's wrong with now?" the crew said in unison. "Let's get moving."

This from a crew that was working 10-and-1/2-hour shifts 6 days a week.

"We did it on Saturday night, and it worked," Matthews said.

"We were blessed with a dedicated crew," Atkinson said, confirming the obvious.

The reasons for this dedication were various. First, all the crew members liked the work and they liked working together. Four crews stayed together throughout the project. Most of them were from upstate New York,

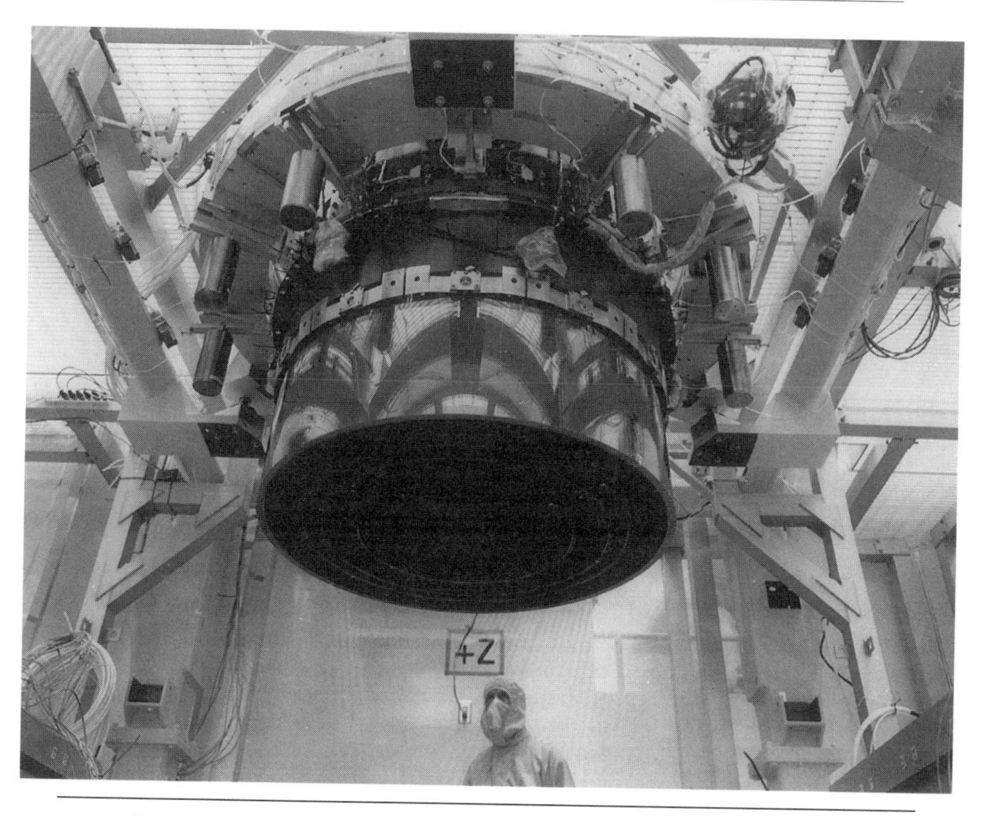

Aligning the paraboloid mirrors of the High-Resolution Mirror Assembly. (Eastman Kodak Corporation.)

and most had worked as toolmakers before coming to the AXAF project. They were used to precision work, though not on this level, and certainly not accompanied by this much stress.

To minimize the effects of gravity, the assembly of the mirrors was done vertically in a 60-foot-high tower. The fragile mirrors were inserted carefully into the support structure, the smallest first. Each mirror was aligned with a specially built laser alignment system, tacked, and then bonded into place. The most difficult mirrors were the four hyperboloid-shaped barrels, or H-mirrors, as they were called. They were at the bottom of the vertical assembly and had to be lowered into the structure.

"We had to hang them with glue and tension," Matthews said. "It could make people a little squeamish, with a multimillion-dollar piece of glass

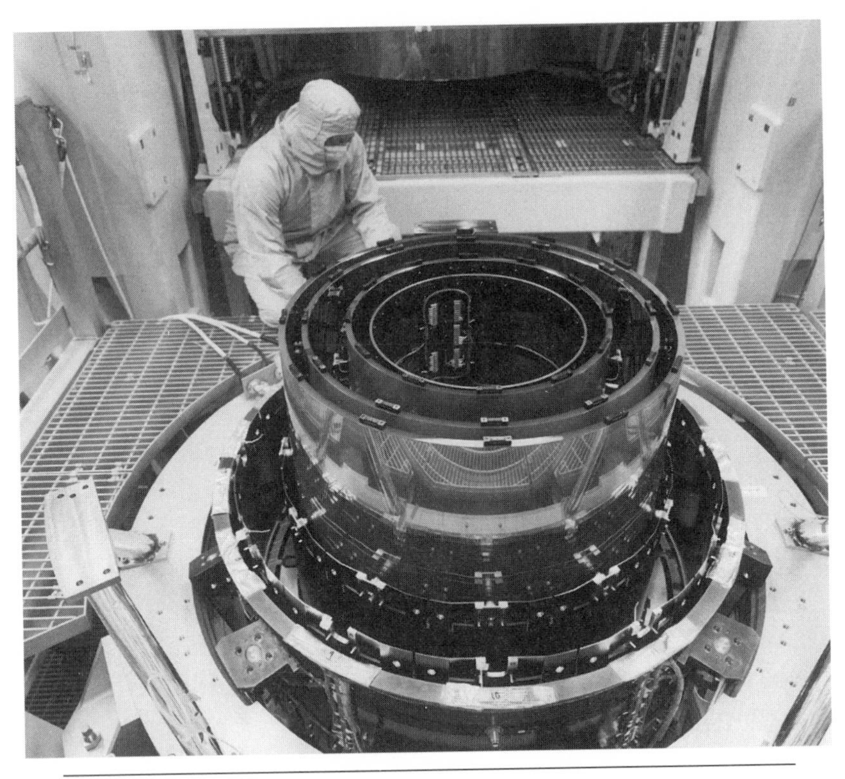

The High-Resolution Mirror Assembly. (Eastman Kodak Corporation.)

hanging there . . . We had tested it on an aluminum mirror, but I got nervous. I wanted to make sure it was going to hold, so we jury-rigged some 2 × 4's and we doubled and tripled and quadrupled the weight on some out there, and it was hanging out there with the real mirror, and I said, 'If those start to fall off, we're going to put the real mirror down quickly. They never fell off.'"

Nevertheless, Atkinson said, "When they were picking up the mirror and it got 10 feet over our head, we got nervous."

"I don't watch when they do that," Matthews said. "I stay in my office."

How did they handle the tension and the togetherness under such stressful conditions? When we talked to one crew shortly after they got off the 4:30 P.M. shift, they appeared remarkably cheerful and relaxed.

"We go out and play golf together," Mark Loucka told us, "especially when we work the night shift and get off at 6:30 in the morning. We don't have any problems getting on the course at that hour."

Another reason for their dedication was mutual respect. "They [the managers] respect us," Femiano said. "They were receptive to feedback from the crew, as when Ed [Swigonski] suggested the tack bonding."

"Most of this is new work," Loucka added. "We like it because it's different."

"Yeah, it's much better than building cars," Dave Sime said. "It gives me great satisfaction to be associated with a program of national significance."

Kathy Rapp, Kodak's manager of the High-Resolution Mirror Assembly (HRMA), understood that sentiment. "I was working at Hughes Danbury when the Hubble was launched," she said. "I saw grown men crying in the cafeteria when it was launched, because they were so proud of what they had done. I always wanted to feel that strongly about my work. Now I do."

Rapp, a mechanical engineering graduate from the University of Illinois, had climbed steadily upward in the management ranks as she moved from one high-tech company to another for 16 years before coming to Kodak. This experience proved especially valuable on the AXAF project with its web of interlocking authorities and responsibilities. Kodak was a subcontractor for putting together an array of mirrors built by Hughes Danbury for TRW, the prime contractor, which worked for NASA's Marshall Space Flight Center. All these groups, plus the scientific mission support team from Harvard-Smithsonian, had a say at one time or another in how the mirror assembly was done. Add to this the need to shift personnel from one job to another and to keep up with the schedule, and the potential for serious breakdowns in communication becomes clear.

One of Rapp's most useful skills, she learned, was her ability to translate from one corporate language to another.

"When you move around a lot, you notice a lot of cultural differences," she said. "There are also a lot of regional cultural differences. At TRW, they will usually tell you what they feel, but it will be more "touchy feely," more California. If you are quiet, it is assumed that you agree. At Kodak, it is harder to know how they feel. For example, if people are quiet, it usually means they disagree."

The contrasting responses sometimes caused misunderstandings on critical issues. "For example, someone came from TRW and presented a plan. Everybody around the table was quiet. The people at TRW said to each other, 'Whew, we can go ahead, they approve.' Meanwhile everybody at Kodak said 'Whew, we didn't have to agree to that!'"

"You could sometimes walk out of the meeting," she went on, "and everyone was cheerful, thinking that they all agreed, but in fact they all disagreed!"

"I am pretty good at translation. If TRW says one thing, I can translate it into Kodak."

When we visited Kodak just before Labor Day in 1996, the need for good communication was paramount, because the members of the Kodak team were running out of time. The slack that they had generously ceded to Hughes Danbury to allow for finer polishing of the mirrors had evaporated under the heat of severe technical challenges. At that point they were 57 days behind schedule.

"It's the 80/20 rule," Rapp explained. "It takes 20 percent of the effort to get 80 percent of the work done, then it takes 80 percent of the effort to get the remaining 20 percent done. That's a real big fear. Now we have to really be done. We really have to finish the wiring, and get the insulation blanket on. All that has been deferred has to be done."

At the status report meeting Jeff Wynn told his staff that Labor Day would be just that, a day of labor. No one complained.

"I'm amazed at the willingness to come in at the crazy hours and crazy shifts that we require," Wynn said.

John Spina, deputy program manager, agreed. "A lot of people here have given up a significant part of their lives for the last 2 years. We've had 10 beautiful weekends in a row–something you don't see in Rochester that often—and these guys don't know it. I try not to bring it up. They're here day and night. They're here 7 days a week. It's a unique opportunity to work on a national science program . . . I've tried to communicate that to them. Most of them have come to that conclusion themselves."

One of the problems Kodak faced was the ancient *Tao Te Ching* curse of "ten thousand things," from electrical grounding of the thermal insulation blanket to a myriad of mechanical devices that had to work perfectly, mostly related ultimately to the desire to focus high-energy X-rays with an accuracy never before achieved. Because of the small grazing angles of reflection needed to focus X-rays, Kodak not only had to align the mirrors to an unprecedented precision, but had to keep them 100 times cleaner than the Hubble mirrors for a period of 3 or 4 years.

"Workers could wear no fingernail polish, no make-up, because flecks and oils could get on the optics," Rapp said.

"We have made a significant investment in a clean room tower," Spina acknowledged. "You could do brain surgery in there. It is a hundred times cleaner than an operating room. We have to clean everything in there for a week. One fingerprint would violate the requirements. It slows you way down . . . It pervades everything we do. We have to keep it clean for 3 or 4 years."

Not only did they have to keep the mirrors clean, they had to keep everything that touched the mirrors clean. For example, the contamination covers that protected each end of the mirror assembly once the testing was complete had to fit tightly so as to keep out contaminants, but not so tightly that they would not open when the time came to start observations.

"We have to make sure that the gaskets on the covers do not stick," said Keith Havey, who had responsibility at Kodak for the covers. "We have been through three different seal designs. We started with something based on expanded Teflon. That has been modified to hard plastic . . . The gasket must give way easily. It must not stick. It must be kept clean, or it will stick."

You get the idea from Havey's comments that he was more than a little worried that the covers might stick, which is understandable. If only one of the covers fails, there is no mission, which is why Kodak built in multiple redundancy. There are four independent springs and a kickout spring in case they don't work.

"I'm going to celebrate when those covers go off," Havey said. "Not until."

Failure of a cover on the mirror to open was an obvious way that the mission could fail. The more subtle modes of failure were what worried Tom Casey, the lead systems analyst for AXAF at Kodak.

"I've found out that the things that we don't have experience with are the toughest," he said.

Things such as how glue dries.

"This was probably the most significant problem in recent times," Casey said. "Tests showed that the epoxy we were using [Heisol 93-13], would meet the requirement. We had 10 years or more experience with that epoxy, and its stability."

It turned out that while the epoxy's stability was correctly known, the way in which it shrank was not calculated correctly. As he did so many times in the program, Harvard-Smithsonian's Lester Cohen identified the problem.

"As the epoxy dried, it did not shrink the same in all three directions," Cohen explained. "This caused the mirrors to move."

The amount of movement, Casey said, "would have put the assembly out of spec, and we would have had to live with a mission that delivered a lot less science."

One possible solution was to take as much moisture as possible out of the epoxy ahead of time. That would reduce its shrinkage considerably and presumably reduce any shifting of the mirror alignment. Kodak contacted people who worked for the manufacturer of the epoxy and asked if that procedure would work.

"They didn't know," Casey said. "That meant that we had to go out and try it, and of course, working with NASA you have to know your processes right down to every little detail. If you change the process, then all bets are off. You have to prove that it won't change the strength and stability characteristics, so that the bond won't break on launch."

Kodak embarked on a qualification program for the epoxy and showed that it did have adequate strength, and much less shrinkage, so they could build the mirror assembly and meet the specifications.

"That was a real nail-biter," Casey said, "because if we had not been able to figure out a way to fix that problem, I think that the program would have gone on, but we would not have been able to achieve 1 arc second resolution by any stretch of the imagination. It would have been 2 or 3. Just that one error would have blown everything else out of the water. It would have doubled or tripled the resolution. It took from late fall of last year to spring of this year [1996] to get that under control."

The last major, most incredible problem that the Kodak team had to solve was even less ponderable than glue. It was air. The body heat of the technicians changed the air. Opening the door changed the air. Turning the light off changed the air.

"We didn't understand the extreme temperature sensitivity," Gary Matthews said. "We tried modeling it, but never could reproduce it. We tried people tests. We would have one person stand beside it [the mirror assembly] for 1 hour. Then three people. We never understood it."

In August of 1996, they were forced to acknowledge that it could be a serious problem. They were having trouble getting the tight, ten-thousandth-of-an-inch focus they wanted on one of the inner mirrors. Was it a problem with the epoxy again? Or the apparatus that was used to offload the effect of gravity? Or the worst-case scenario of a mirror ground to the wrong shape?

"We were stumped," said Atkinson. "Then, we turned the lights off to make a measurement with another instrument, and forgot to turn them back on."

"These are 40-watt fluorescent lights, like [those] in a normal office, 10 feet away," Matthews added.

"With the lights off the effect went away," Atkinson said.

Further investigation—flipping the lights on and off, and making repeated measurements—showed that the fluorescent lights were the problem. Calculations showed that the lights were heating the inner support structure, which in turn heated the surrounding air by 3/100ths of a degree.

"The mirage effect," Mark Freeman explained.

The tiny change in the temperature of the air in the mirror assembly had caused a minuscule bend in the path of the laser test beam, in the same way that hot air near the desert floor or a hot blacktop road bends light and produces an illusion. From that point on they left the lights off and aligned the mirrors with the lights out.

On November 30, 1996, Kodak shipped the mirror assembly to the Marshall Space Flight Center. The shipment was behind schedule, but not so far behind as to impact the overall program schedule. It had been a grueling 5 years with little time off for the workers, counting the time they had worked on the mirror challenge, but few regretted it.

"It's been a long, long road and quite a ride for me," said Casey. "It's been a good ride. I'll miss it when it's gone."

"You won't see the likes of AXAF again in my lifetime," Spina said, ". . . not for another 30 or 40 years. These will be the only high-resolution mirrors for many, many years."

"The entire community has pulled together," Wynn noted. "If we had a problem, NASA, SAO, or Marshall was on the phone, holding telecons on the weekend. Leon [van Speybroeck] and Marty [Weisskopf] came often. This total community involvement makes it more fun than trying to do it yourself. People willingly gave up their leisure time, whether they are at MSFC [Marshall Space Flight Center], NASA, SAO, Kodak, or TRW . . . It's going to be the end of an era when they get it done, and the beginning of a new era when the data starts coming in . . . it's fascinating to be part of that."

20

Calibration

ON NOVEMBER 30, 1996, the AXAF mirror assembly was rolled gingerly off a modified C5A transport plane that had brought it from Rochester, New York, to the Marshall Space Flight Center in Huntsville, Alabama. The AXAF team was confident that these were far and away the best X-ray mirrors ever made. However, mistakes do slip by even the most watchful eyes, as experiences with the Hubble Space Telescope had made painfully clear a few years earlier. In science, it is proof, not confidence, that ultimately matters. Over the next 6 months the AXAF mirrors and instruments would be thoroughly tested in Marshall's mammoth X-ray Calibration Facility to determine just how good they were.

Apart from the obvious need to catch any glaring errors, the need for testing was especially critical for AXAF, precisely because it was expected to be so good. If the images really did turn out to be 50 times better than those made by previous X-ray telescopes, then astronomers would be seeing things they had never seen before. Optical and radio images could be used as a guide, but the history of X-ray astronomy showed convincingly that an X-ray telescope reveals features that no other telescope can detect. The response of the mirrors to X-rays of a wide range of energies, as well as the off-axis, or peripheral vision, of the mirrors, had to be tested—calibrated—under carefully controlled conditions. Otherwise, an artifact of the telescope's response to X-rays might be mistaken for a scientific discovery. For example, you calibrate your sunglasses when you look at a scene with them on, and then without them, to see how much "rosy" distortion they might be introducing into your perception. Or you calibrate the odometer in your car when you check it against a carefully measured set of mileage markers along the highway.

Months of preparation went into the calibration. Under the overall direction of TRW, the Marshall AXAF team completed the construction of the facility, which had been begun in preparation for the mirror challenge 4 years earlier. The Harvard-Smithsonian group built detectors to test the telescope and made a list of desired tests, and the Marshall team developed a detailed schedule for each of more than 250,000 different tests.

Around the third week of December, the eager scientists got their first look at how well the mirrors could focus. The largest mirror pair had tested out very well with X-rays in the mirror challenge, and Kodak had tested the assembly of four mirror pairs in visible light, but the X-ray Calibration Facility, with its third-of-a-mile-long tube to simulate X-rays from a distant source and a huge vacuum chamber to simulate the conditions of space, was the only place in the world to test the AXAF mirrors with X-rays.

"Around the third week of December," Harvey Tananbaum recalled, "we pumped the air out of the vacuum chamber and waited a few hours until we were sure all the contaminants had been cleared out of the chamber. By this time it was well after midnight."

Martin Weisskopf and two other members of the Marshall AXAF team, Steve O'Dell and Jeff Kolodziejczak, were there. Leon van Speybroeck and Christine Jones, head of the Harvard-Smithsonian calibration team, were also there to watch as the command was given to open the covers on the mirrors and expose them to X-rays for the first time.

"We adjusted the focus and there it was," Tananbaum related. "A nice sharp image, just like we had predicted. It was a very exciting time!"

Over the next few weeks the excitement would give way to frustration and anxiety. The scientists encountered problems with the mechanical devices used to move pieces of test equipment and instruments in and out of position.

"The actuators kept going wrong," Jones said. "The shutters kept getting stuck . . . we often couldn't get the data processed in a timely fashion."

To Jones, who had worked on the calibration team for the Einstein Observatory almost two decades earlier, these problems were understandable, because "we did a lot more stuff than with Einstein . . . Besides, the measurements were so good that it made it easier to deal with the difficulties. We showed that the mirrors are good, and the instruments are good."

Early in the testing the team discovered a significant problem. While analyzing a particularly perplexing set of data, the Harvard-Smithsonian scientist Terry Gaetz discovered that the mirrors were not centered properly. This misalignment can cause the upper part of the mirrors to have 10 percent more efficiency than the lower part.

"Because of calibration, we now understand this effect and can correct for it," Tananbaum said. "Otherwise, every image we made would have been off by a few percent and we wouldn't have known it."

Dan Schwartz, the science operations group leader at Harvard-Smithsonian, agreed. "Existing X-ray mirrors in orbit may have this problem and we would never know. It was a triumph to discover this."

The testing went on, 24 hours a day, 7 days a week. Exhausting, exacting measurements, through Christmas, through New Year's, and on through the winter. The senior scientists did not take executive privilege, but worked right alongside the rest of the team. Tananbaum, for example, worked at the Harvard-Smithsonian Observatory Monday through Thursday, caught a plane to Huntsville in time for the three-to-midnight shift, worked that shift through Sunday night, and and then flew back to Cambridge to be ready for work Monday morning.

"It was tiring," Tananbaum admitted, "but it was an exhilarating and energizing process, seeing it all come together and actually work, so you didn't realize how hard you were working."

Martin Weisskopf felt the same way.

"It was a bonding experience for 200 or 300 scientists."

It was also a near-death experience for Weisskopf. As AXAF project scientist and the lead scientist from the Marshall Space Flight Center, Weisskopf had a dual responsibility that included getting the calibration facility ready and keeping it running throughout the calibration period. A tense, chain-smoking, competitive man who could be alternately charming and cranky, Weisskopf liked to relax by playing basketball before his three-to-midnight shift.

"I had left the gym, and was driving home to get ready for work," he said. "I didn't feel very good—hadn't felt good for several days, actually—but I thought maybe I had pulled a muscle playing basketball. I kept driving, but I began to feel much worse. I thought, 'Maybe I should stop driving, and walk around a little.' The next thing I knew I was in the hospital."

Fortunately for Weisskopf, a medical technician who happened to be driving behind him saw him collapse. Weisskopf had suffered a heart attack. The technician immediately called 911 and got Weisskopf to the hospital.

"Thank goodness it happened right after the testing was over, so it didn't slow the testing down," Weisskopf told us with a laugh a year after the episode. Giving credit to his wife, Mary Ellen, Weisskopf quit smoking, changed his diet, and made a full recovery. He has resumed his daily basketball game and assures us that he is competitive, even though most of the players are much younger.

"I don't have to win," he said. "But I do like to play the game."

21

The Scientific Instruments

THE EXHAUSTIVE and exhausting test program at the Marshall Space Flight Center confirmed that AXAF's mirrors were, as advertised, capable of producing images 50 times sharper than those produced by any previous X-ray telescope. But as even an amateur photographer knows, a camera with a flawless and powerful lens is of little use without film that is able to capture the quality of the image. Astronomers now use electronic detectors for almost all of their research. They are more sensitive than film, and for space research they are a necessity. The data are collected, stored onboard in a computer, and then transmitted as a radio signal to a receiver on the ground.

The two detectors selected for AXAF were chosen to make the most of the focusing power of the mirrors. The High-Resolution Camera (HRC) has the closest match of imaging capability to the focusing power of the mirrors. The HRC can make images that reveal detail as small as half an arc second. This is equivalent to the ability to read a stop sign at a distance of 12 miles. The HRC is especially useful for making images, as well as for identifying very faint sources.

The Advanced CCD Imaging Spectrometer (ACIS) is the other focal plane instrument. As the name suggests, this instrument is an array of charged coupled devices (CCDs), which are sophisticated versions of the relatively crude CCDs used in camcorders. This instrument is especially useful because it can make X-ray images and at the same time, measure the energy of each incoming X-ray. It is the instrument of choice for studying temperature variations across X-ray sources, such as vast clouds of hot gas in intergalactic space, and for tracing the heavy elements in remnants of supernova explosions.

Additional instruments allow astronomers to obtain even more detailed information about the energy of incoming X-rays. These are assemblies of finely constructed gold gratings that are attached to the rear of the mirrors, and can be swung into the path of the X-rays. In much the same way as a raindrop breaks sunlight into the colors of the rainbow, these gratings bend or diffract the reflected X-rays according to their energy. High-energy X-rays are bent less than low-energy X-rays, so they strike the detector at different places. A precise measurement of the position where the X-ray hits the detector gives an exact measurement of the energy of the X-ray. The grating literally sorts the X-rays according to their energy.

By observing a cosmic source for a long time with the transmission gratings in place, X-ray astronomers can construct a graph that tells with great accuracy how many X-rays the source radiates at each energy. This graph, or energy spectrum as it is called, is one of the most important pieces of information that an astronomer needs to decipher what is going on in the distant object that is producing the X-rays. Beautiful images are the art of astronomy. Energy spectra are the business.

With energy spectra astrophysicists can do what the French philosopher Auguste Comte cited as the epitome of the impossible: they can determine what stars are made of. They can do so because each element has its own unique structure, which is the same wherever you find it in the universe. A hydrogen atom extracted from the water in your bathtub is the same as one in the sun, or in a star a thousand light years away, or in a galaxy 10 billion light years from Earth. The same goes for an oxygen atom, or any other elemental atom or ion formed by stripping one or more of the electrons from the atom.

The orbits of electrons in an atom are strictly regulated by the rules of quantum theory. These orbits, or more accurately, quantum states, are separated by a specific amount of energy, just as stairs are separated by a specific height. The movement of electrons between quantum states in an atom can be likened to movement up and down stair steps. If you wish to move up or down the stairs, you must move from one step to another one. You cannot move to a position between the steps. You might do two steps at once, or even three, but you couldn't do two and a half steps. Likewise, an electron in an atom can move from one specific quantum state to another one in whole steps.

When an electron jumps down from one quantum state to a lower one, the energy of the electron has decreased by an amount of energy equal to the difference of energies in the quantum states. This energy is radiated away in the form of a photon of a specific energy—equivalent to the height of the step in our stair step analogy. Careful study of the energies of the photons given off by an atom, say a hydrogen atom, reveals the blueprint for the quantum states, or energy spectrum, for that atom. Knowing this spectrum, astronomers can look for it in the radiation from stars and gas, and determine the amount of hydrogen in any given object. They can do the same with any other element.

The energy spectrum for a star or a diffuse cosmic gas cloud is at first glance a chaotic graph of peaks and valleys that looks like an EKG of an irregular heartbeat or a summary of a turbulent week on the stock market. But the trained eye can immediately pick patterns that tell which elements are present, and approximately how hot the star or cloud is. Further analysis can usually yield the amount of each element. In this way, astronomers have determined that the stars are mostly made of hydrogen, with a mixture of helium and traces of heavier elements such as carbon, nitrogen, oxygen, and so on.

By observing subtle changes in the energy spectrum, astronomers can also determine how the atoms are moving, whether in a random hurly-burly way, as in a turbulent gas cloud, or in an orderly way, as when one star orbits another. When astronomers announce that they have discovered a new planet around a star or matter falling into a black hole, or that they have measured the amount of deuterium in the universe, they have relied on detailed studies of an energy spectrum for their findings. With a few exceptions, virtually every research paper that discusses the observation of a cosmic object makes use of an energy spectrum directly, or refers to one.

Two factors have hampered the exploration of the wonders of the high-energy universe: the lack of a telescope with the ability to make a sharp image (angular resolution) comparable to optical telescopes, and the inability to make precise measurements of the energy spectrum (energy resolution). AXAF would be the first X-ray observatory to have both such a telescope and the ability to measure the energy spectrum precisely.

Two instruments aboard AXAF would be dedicated to making high-resolution energy spectra. These are the High-Energy Transmission Grating Spectrometer (HETGS), made by a group at MIT under the direction of Claude Canizares, and the Low-Energy Transmission Grating Spectrometer (LETGS), made by a collaboration of the Space Research Organization of the Netherlands (SRON) and the Max Planck Institute for Astrophysics in Germany under the direction of Albert Brinkman.

The high-energy grating has an array of 1-inch-square facets composed of fine bars mounted on rings matched to the AXAF mirrors. For an X-ray grating the challenge is to make exceedingly fine and precisely spaced bars so that the X-rays will be spread out to reveal their spectrum. Seen through a microscope, the gratings would look like a picket fence made of gold, with a separation between the pickets less than a hundred thousandth of an inch, less than the wavelength of light. It would take hundreds of bars to equal the thickness of a sheet of paper. The bars are supported on the facets by plastic membranes that are as thin as a soap bubble, yet can withstand the trauma of a shuttle launch.

Originally, the Canizares group planned to use X-ray lithography to make the more than 300 facets. In a process similar to stenciling, a mask is laid over a substrate of material, and then a high-power X-ray beam is used to "burn out" the gratings. The same mask is used many times to replicate the same grating pattern on many facets. MIT had a lower-power X-ray machine in the laboratory, but it took 48 hours to make a single grating and the MIT scientists needed hundreds. They needed a high-intensity X-ray machine. Fortunately, or so it seemed, they located a small start-up company in New Hampshire that was in the process of building high-intensity X-ray machines for industrial application. They ordered one, at a price of several million dollars, and made plans to begin the process.

Less than a month before a critical design review, Canizares received the chilling news that the company had gone bankrupt and the president had committed suicide. There would be no X-ray machine.

"Suddenly we were stuck without a key piece of equipment," Canizares said. Mark Schattenburg, who started out on the project as a graduate student and eventually became the instrument scientist, came to the rescue. Schattenburg was in charge of making the mask.

"Mark by this time had done so well at making the mask," Canizares said, "that he proposed we try to make hundreds of masks and fly those.

The two big questions were, could we make them thick enough to have high efficiency, and could we make them all the same? So in a frenzy he set out to prove we could meet those two hurdles."

Working nights and weekends in his home workshop, Schattenburg showed that his process could meet both the specifications and the schedule. Within 3 weeks of the calamity, Canizares was able to take a working prototype to a meeting at the Marshall Space Flight Center. Before Schattenburg and his colleagues finished the final product, their work led to several industrial spinoffs in the area of nano-technology with applications to high-speed electronics, lasers, and flat-panel displays.

The High-Energy Transmission Grating Spectrometer consists of two sets of gratings mounted on the same structure. One set, the medium-energy grating, intercepts X-rays reflected from the outer two pairs of mirrors, and it gives the best performance for medium-energy X-rays. The high-energy grating intercepts X-rays reflected from the two inner mirror pairs, and is optimized for high-energy X-rays with wavelengths of less than 2 nanometers (1 nanometer = 40 billionths of an inch). Such X-rays are produced primarily in gases with temperatures greater than about 10 million degrees. Gas this hot is found in the hot atmospheres of some peculiar stars, around black holes and neutron stars, in shock waves produced by supernovas, in the space between stars in galaxies, and in galaxy clusters.

The Low-Energy Transmission Grating Spectrometer has a structure and design similar to its high-energy counterpart, except that it uses fine, closely spaced gold wires rather than gold bars. The spacing of the wires is arranged so that it can most effectively measure the energy of low-energy X-rays, which are primarily produced by gases with temperatures of less than about 10 million degrees. These include the hot upper atmospheres, or coronas, of most stars, white dwarf atmospheres, and possibly the cool central region of some galaxy clusters. The LETGS will also be useful for studying cool extended gas clouds around bright X-ray sources associated with giant black holes.

Albert Brinkman of the Space Research Organization of the Netherlands is the principal investigator for the low-energy transmission grating. He and his team had extensive experience building similar instruments for the Einstein and the EXOSAT X-ray observatories. During the long delay in getting AXAF funded, two key members of his team retired.

High-energy (outermost with disks) and low-energy (innermost with circles) transmission gratings. (TRW.)

Rather than embark on the risky course of training new people in an esoteric technology, Brinkman chose the wise course of inviting his former competitors to join in a collaboration.

He approached Peter Predehl of the Max Planck Institute for Extraterrestrial Physics in Garching, Germany, the leader of the team that had 20 years' experience working on X-ray diffraction gratings. Predehl's group immediately enlisted the services of the German company Johannes Heidenhain GmbH in Trauenreut, a small town in Bavaria. Johannes Heidenhain, the founder, was known for his willingness to have his company take on projects for research groups that required new technologies in the field of high precision micro-mechanics. His philosophy, which has been borne out by the company's success, is that the company could keep its competitive edge by developing new manufacturing techniques.

In the course of designing the low-energy grating, Predehl soon hit upon a major logistical problem. The gratings would have about 3,000 facets, which had to be manufactured and tested to see if they met the specifications. After testing a few samples, he and his colleagues realized that testing all 3,000 facets would occupy all the available X-ray test facilities at the Max Planck Institute for many years. One of Predehl's doctoral students, Hans Lochbihler, came across work by French scientists on "stealth" equipment—hangars that are invisible to radar—that he thought might help solve the problem. By adapting this work, Predehl and Lochbihler showed that they could do most of the measuring with visible light—for which ample facilities were readily available—and use X-ray diffraction theory to predict the X-ray performance.

An additional benefit was that this procedure made the work more interesting.

"It was much more fun to develop a new method than [to follow] the boring approach of simply measuring all gratings individually but in an identical manner," said Predehl.

Another serious problem was the sensitivity of the thin wire gratings to noise. Gold wires half a micrometer wide (less than 1 percent of the width of a human hair) are stretched like a delicate cobweb under high tension. When subjected to the sounds typical of a shuttle launch, they disintegrated. A Max Planck Institute engineer, Gunther Kettenring, made extensive calculations and devised a new design for the support structures that worked. The group was able to deliver the gratings on time to the X-ray Calibration Facility for testing.

The collaboration between Brinkman's group in Holland and Predehl's group in Germany went well from start to finish.

"It was a beautiful and fruitful collaboration that lasted over 10 years," said Predehl. From that point of view, Predehl continued, "I never had such a smooth project!"

The transmission grating must be used in conjunction with one of the X-ray cameras. The high-energy transmission grating is designed for use primarily with the ACIS, and the low-energy grating with the HRC. Although the precise measurement of X-ray spectra is rich with valuable information, astronomers often choose to make measurements without the gratings. The primary reason for this is practical. The gratings disperse a large fraction, typically more than 90 percent, of the incoming X-rays. Since one of the main reasons for building a revolutionary X-ray telescope like AXAF was to study sources too faint to be observed by previous telescopes, astronomers often cannot afford to lose so many photons, so they will observe with the ACIS or the HRC alone.

The HRC, or High-Resolution Camera, has an impressive heritage. It is the descendant of the high-resolution imagers used on the Einstein X-ray Observatory and the Roentgensatellite, or Rosat Observatory. Steve Murray of Harvard-Smithsonian, the principal investigator for the HRC, played a major role in the design of the Einstein imager and a leading role in building the imager for Rosat.

Murray became involved with the AXAF program in 1982, when the first call went out for investigations of possible instruments. After some discussion among the group at Harvard-Smithsonian, it was decided that scientists would not employ the tactic they had used successfully for the Einstein Observatory, namely to form a consortium that would be responsible for all the scientific instruments on the observatory as a group. Rather, each scientist would propose on his or her own. In 1984 Murray submitted a proposal for an updated version of the detector he had helped design and build for the Einstein Observatory. In 1985 he was informed that his proposal had been selected, along with five others.

"HRC [the High-Resolution Camera] was chosen because it was a known technology, and was pushing the technology in an incremental way," Murray said. "They [the members of the NASA selection panel] felt very comfortable with that."

The primary components of the HRC are two micro-channel plates, postcard-sized clusters of 69 million tiny lead-oxide glass tubes that are about 10 microns (1/2,500th inch, or about 1/6th the diameter of a human hair) in diameter and 1.2 millimeters (1/20th inch) long. The tubes have a special coating that causes electrons to be released when the tubes are struck by X-rays. These electrons are accelerated down the tube by a high voltage, releasing more electrons as they bounce off the sides of the tube. By the time they leave the end of the tube, they have created a cloud of 30 million electrons. A crossed grid of wires detects this electronic signal and allows the position of the original X-ray to be determined with high precision. With this information, astronomers can construct a finely detailed map of a cosmic X-ray source.

In the past, one of the stickiest problems encountered in building the High-Resolution Imager for the Einstein Observatory had been to fabricate the crossed grid of wires. The scientists and technicians had solved this problem not by adopting the sophisticated technology used by the semiconductor industry but by using a simple mechanical winding tool. By the mid-1990s, the semiconductor industry had undergone explosive growth, in its technology as well as its profits. It seemed obvious that this time around, they would make use of the new expertise.

Murray is an instrument builder in the classic sense. He likes the lab; he likes to do precision work; he even likes hanging wallpaper, because it demands exactness.

"I think of myself as a Swiss watchmaker," he said.

A Swiss watchmaker who is as comfortable with a computer as with a watchmaker's screwdriver, and who understands just how hostile and unforgiving the environment of space can be.

"We [the HRC team] did 2 years of research on other technologies to do this, and ultimately decided to go back to the tried and true technology. Sure, Intel or some other company could have built the read-out device for us, but it would have cost millions, since we only wanted two of them. Even then, we wouldn't have been sure that it would work once it was in space. It turns out that it is cheaper and more reliable to do it the old-fashioned way because you have more control and you minimize the risk."

They used the same device they had used 20 years ago to fabricate the crossed-wire assembly for the Einstein Observatory. And what's more important, most of the same crack scientific and engineering team—with Gerry Austin, John Chapell, Dick Goddard, and Martin Zombeck, who had been with him for years—was still intact. In August of 1996, Murray thought they were in good shape to meet the scheduled testing that would begin that winter at NASA's Marshall Space Flight Center.

"It's an amazing thing," he told us then, "it's all coming together. We have all the necessary parts on the table, and we've added new features to make our detector work better with the transmission gratings—all the main parts are done . . . If we get lucky and get through vibration and thermal vacuum early, we will have 5 weeks to calibrate."

But you can't count on getting lucky when you are building complex, sensitive instruments that must survive the tumult of launch and then a decade in space.

"We assembled the first build-up of the flight detector," he said later. "And we started testing. Within hours we had problems—irregular background, increased background."

The HRC team members kept testing, hoping that their test protocol was the problem. The poor performance persisted.

"We didn't know what was going on."

What they did know was that they had less than 2 months to fix it, or they would fall behind schedule. They decided to take the problematic unit apart, and look for dirt or any other problems. This had to be done with great care, lest they damage other components in the process.

"Each cycle took 5 days to a week," Murray said. "After three or four attempts at fixing it this way, we concluded we were on the wrong track."

They tried another approach. They mixed and matched test parts and flight parts to see if they could get a configuration that was stable, and thereby identify the faulty part.

"We did that for a couple of weeks. Again, we got no conclusive results. We had all these tests, we had all these ideas—was it a problem with the insulation in the frame? Was it a static charge? Outgassing?"

Because so much electronics must be packed into such small spaces, a major worry in all spacecraft components is static discharge, the jumping of electric charge across a narrow gap between two elements of a circuit.

This static discharge, or breakdown, or corona, as it is variously called, can produce unwanted noise, unexpected read-outs, or in the worst case ruin an experiment. Murray's team looked at the design tolerances, and modified them wherever possible to make the spacing between components larger.

"In the end we decided that it was a high-voltage problem," he said. "We could see excessive current being drawn out of the high-voltage power supply when we got the noise problem."

After 3 months the situation was critical. Now they were behind schedule and with each passing day they were falling further behind. Then they had a breakthrough.

"We put the system together and let it run for a whole day in its bad condition. Then we took it apart and inspected it visually. We didn't find anything, until almost by accident, but more out of desperation than accident, we took apart a test fixture that held the unit."

On the back side of this base plate, they found what they were looking for—small burn marks, evidence of high-voltage discharge. They immediately understood the problem.

"It's embarrassing to say, that it was just a set of four screws . . . Normally, in the flight unit they would be insulated, but in the test set-up they weren't."

As a result, a 4,000-volt potential difference was created that was causing both the discharge and the background noise.

"It was a totally trivial thing to fix, and nothing to do with any of the other things we had investigated. However, all the redesign we did will probably help us in the long run."

By now it was September of 1996, and they still had to finish putting all the parts together to make the completed instrument. They had to inspect each printed circuit board, test it with high voltage to make sure there were no loose wires, and test it again at operating voltages to make sure it performed according to specifications. This process took 8 to 10 hours for each circuit board. The HRC team worked 90-hour weeks in an effort to make up for the lost time. It became clear they would miss their November delivery date. They had to rush, but they knew that they couldn't rush. Each of a thousand wires had to be carefully connected. Finally by the end of December they were ready for the final assembly.

The final assembly process required many of the skills of a brain surgeon—nerves of steel, sure hands, strong arms, and extreme confidence that the assembler knew what he was doing. The assembly has to take place in a Class 100 clean room. As a further guard against contamination, the detector was inside a sealed compartment within the clean room. A steady breeze of dry nitrogen gas blew across the detector to keep any particles from settling on it. The assembly of the detector had to be done with a glove box that allowed the assembler to reach into the sealed compartment by inserting his hands into gloves that were connected to the walls.

The entrance to the HRC is covered with an extremely thin film, or window, of aluminum-coated polyimide. The window, which is called a UV-ion shield, is designed to prevent contamination and to absorb ultraviolet radiation and ions without absorbing the X-rays coming into the camera.

"One quick movement of your hand will set up air currents that will break the window. It is very fragile and very awkward to handle."

Only four people were allowed to handle the HRC—Murray, Austin, Chapell, and Almus Kenter. Only two could handle the detector at a time, since the glove box had only two pairs of hands. One of those pairs almost always belonged to Murray. The other was usually manipulated by Austin, Murray's longtime colleague.

"I personally took the attitude that I should work on it most of the time, because I didn't want anyone else to feel the pain if it got broken," Murray said.

One of the key procedures in the assembly was to torque down each screw and then bond it with a little dab of epoxy so that it wouldn't break loose during vibration testing or launch.

"It's unfortunate, but true, that you forget to do something, or something doesn't fit together right, so you have to take one or more of the screws off," Murray recounted, with obvious pain in his voice. "This means you have to break off the glue."

On one such occasion, a little sliver of glue flew up into the air, and landed on one of the thin plastic windows.

"It didn't break it, but it was sitting there and we had to remove it."

They tried very delicately to remove the sliver of glue and watched in horror as a crack in the window spread across the detector. They had to replace the filter, which required a major disassembly of the whole detector—another couple of 15-hour days. Things went well for a few days, then Murray dinged the window on one of the detectors.

Murray was devastated. He knew that he had undone weeks of work. They would have to take the detector apart and change the window. It wasn't even obvious how they would do it. The protocol for assembling the detector was carefully spelled out, but none existed for taking it apart, especially for people working in clean room bunny suits through a glove box. They could easily damage the detector permanently if they made a mistake. Murray spent a day with his team discussing what to do, and trying to come to terms with his mental anguish over his mistake.

"I've let everyone down, and all the work that everyone has done has been shot to hell because I was careless," he told them.

Austin, a gruff, no-nonsense engineer who had worked with Murray on the Einstein and Rosat projects, had no use for such self-flagellation. "Let's stop worrying about that and figure out how to fix it," he said. "Let's get on with it."

"There were no recriminations, no second-guessing," Murray said, "His support was really wonderful and important to me. The same was true of the other team members. They were very supportive."

By the time they got the detector put back together it was January. They still had to put the electronic harness together that would link the detector to the rest of the spacecraft. The mirrors were at the calibration facility, and the time was rapidly approaching to test the High-Resolution Camera.

"Fred Wojtalik, the AXAF program manager, was calling almost every day," Murray said, "telling us that if we didn't get there soon, they couldn't test it, and if they didn't test it, they wouldn't fly it. The pressure was enormous."

The work proceeded without any major problems, but a series of minor problems, plus a delay in the development of the high-voltage power supply, continued to slow them down—mistakes such as a faulty wiring by a color-blind technician who couldn't tell the difference between brown and purple wires.

The High-Resolution Camera. (SAO/CXC.)

Wojtalik continued to issue ultimatums.

"We were told that if we didn't have the detector in the truck by February 22, don't bother. That was our last chance date."

They were able to test the detector by itself to verify that it worked, but had no time to check it with the flight power supply and electronic harness. At 2:00 A.M. in the morning of February 22, they loaded the detector, power supply, and harness onto an Air-Ride van bound for Huntsville. After a few hours of sleep, they were on an airplane to Huntsville to prepare for the final tests with the mirror assembly at the X-ray Calibration Facility. After 6 weeks of testing in Huntsville, and another 5 weeks of testing and trouble-shooting back in Cambridge, the HRC was shipped to Ball Aerospace in Boulder, Colorado, where it would be installed into the science instrument module that would hold both the HRC and the ACIS, the other focal plane instrument.

In the late 1970s highly efficient new solid-state detectors called charged coupled devices, or CCDs, were rapidly becoming the detectors of choice for optical astronomers. Now commonly used in digital cameras and camcorders, CCDs were a product of the silicon revolution in which thin silicon wafers or chips were discovered to be capable of many amazing things, all related to their ability to become conductors with the addition of minute electric voltages. The essence of the operation of a CCD is this: An incident photon produces a shower of electrons that fills up little wells, or pixels, in a silicon chip. By the clever application of voltages across the wafer, these electrons can be counted, and it can be determined precisely where the photon struck the chip. This information can be used to reconstruct an image of the source of the photons. CCDs are, in effect, an extremely efficient type of electronic film that is perfectly suited for research in space, since the electronic signals can be readily converted to radio signals and transmitted to the ground station.

Many X-ray astronomers realized that future X-ray observatories would use CCD detectors. Gordon Garmire was among them, and he was probably also among those who underestimated the difficulty of building a large CCD X-ray detector. Garmire is a veteran X-ray astronomer with a long résumé that stretches back to the pioneer days of the sixties, when he was part of the MIT X-ray astronomy group. He left MIT to go to Caltech, was one of several principal investigators on the first High-Energy Astrophysical Observatory, HEAO-1, which flew in the mid-1970s. In the late 1980s, Garmire began designing an X-ray detector that would use CCDs. When the scientific instruments were selected for AXAF, his proposal for the Advanced CCD Imaging Spectrometer, or ACIS, was one of the winners.

Garmire's instrument survived the restructuring of AXAF in 1992, but left him with some worries. One of these was the high Earth orbit, which would take the satellite through the Van Allen radiation belts for about ten hours each orbit. The detector would also be more vulnerable to solar flares in a high Earth orbit, since it would be above the protective shielding provided by the Earth's magnetic field. Technicians installed protective shielding, but Garmire was still worried about it.

"I am concerned about the sun," Garmire told us. "We could be unlucky and have a lot of solar damage at the beginning of the mission. You just never know. The peak of the solar cycle is around the year 2000."

One previous X-ray astronomy satellite, the European-American collab-

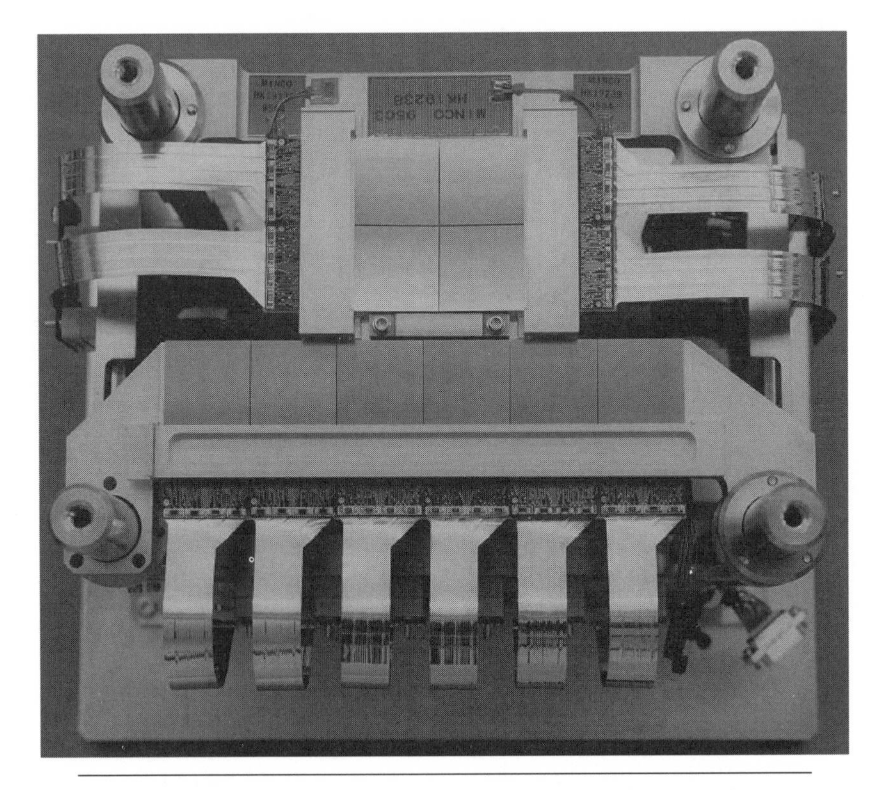

The Advanced CCD Imaging Spectrometer (ACIS). The CCD chips are arranged in imaging (four chips in a square in the upper middle) and spectroscopic (six chips in a row near the center) arrays. The flex prints are the metallic ribbons attached to the CCD chips. The door is not attached in this photo. (Pennsylvania State University/MIT.)

oration EXOSAT, had an orbit similar to the one proposed for AXAF. An analysis of data obtained by EXOSAT between 1983 and 1986 showed that a lot of data were lost because of the background, or static, produced by collisions of the numerous high-energy electrons and protons with the detectors. Garmire added a set of eight magnets to the telescope to act as a "magnetic broom" to sweep the charged particles away from the detectors.

In the late 1980s, Garmire left Caltech to set up what has become a flourishing X-ray astronomy group at Pennsylvania State University. Shortly thereafter, he formed a partnership with MIT and Japanese as-

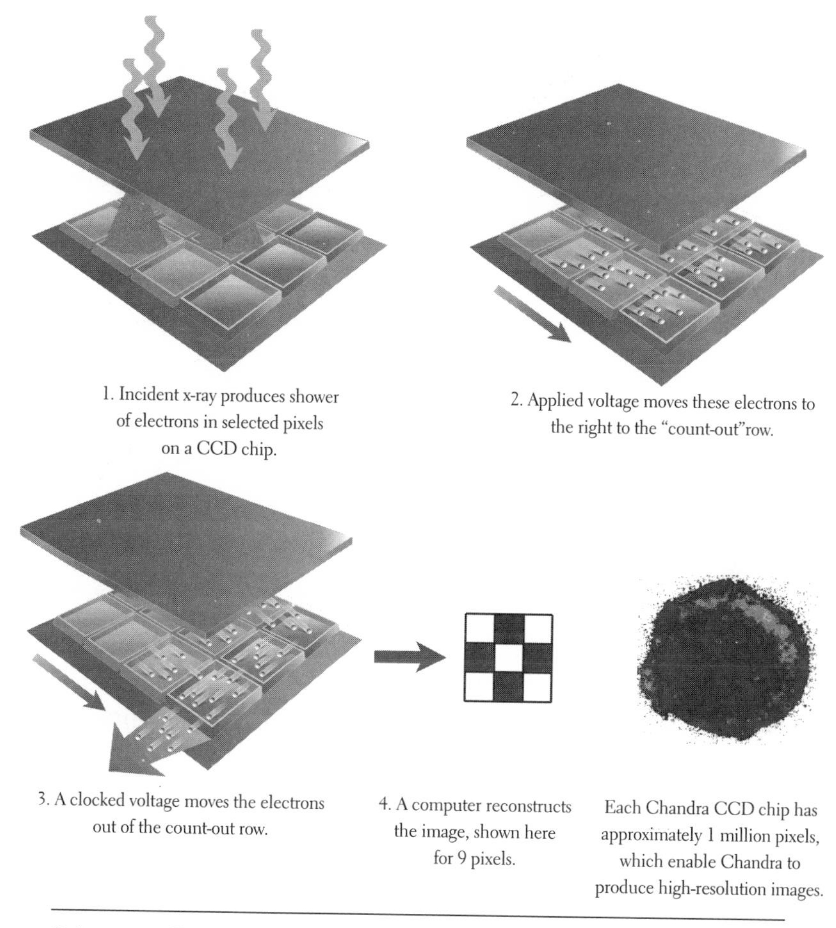

Schematic illustrating the operation of the Advanced CCD Imaging Spectrometer (ACIS). (SAO/CXC/S. Lee.)

tronomers to build an X-ray CCD detector for the Japanese Advanced Satellite for Cosmology and Astrophysics (ASCA).

"It was a distraction from building ACIS," he said, "But it was a proof of concept."

Launched in 1993, ASCA had foil mirrors with low resolving power. Nevertheless, the performance of the CCD detectors showed that CCDs would play an important role in future X-ray astronomy missions for years to come. It also showed that CCDs are not error free and that great care must be exercised to acquire CCD chips of superior quality.

After investigating several suppliers, most of whom either could not exercise the extreme quality control demanded for a long-term X-ray mission or could not guarantee timely delivery, Garmire decided to form a complex partnership. Lincoln Laboratory of Lexington, Massachusetts, would supply the CCDs, Lockheed Martin Aerospace of Denver, Colorado, would build the structure, power supply, and data-processing unit, and an MIT team under the direction of George Ricker would put the instrument together, all under Garmire's supervision.

This collaboration went well until the early 1990s when MIT's excellent but small engineering group became overextended. The members of the group were committed to build a major portion of the Rossi X-ray Timing Explorer (RXTE), a satellite named after MIT's pioneering X-ray astronomer, Bruno Rossi. Since the RXTE was scheduled to launch in late 1995, the MIT group had to pull people off AXAF to work on RXTE.

"We were stuck with a skeleton crew," Garmire said. "The program was slipping and we were headed for disaster."

Canizares, who is also director of MIT's Center for Space Research, acknowledged the problems.

"Building the ACIS was slightly larger than the largest project we should be running through here," he said at the time. "Marshall cut costs, and we had to take on additional work here. This, compounded with the X-ray Timing Explorer, created scheduling problems."

MIT moved one of its best engineers, Bill Mayer, from RXTE to AXAF.

"Mayer and Bob Goeke put the program back on the rails," Garmire said, when we talked to him in May of 1996. "They really came through for us."

Normally a CCD is on a wafer about 300-500 microns thick. The thickness is adjusted by etching it with corrosive acid, and the back side of the chip is heated with a laser to anneal it. In the course of preparing an ultraviolet detector for another satellite, George Ricker discovered that when he flipped a chip over and used it upside down, it became a good detector of low-energy X-rays. These so-called backside-illuminated chips were more desirable than the normal, front-illuminated chips for detecting low-energy X-rays, but had disadvantages. They yielded cruder spectra than the front-illuminated chips, they turned out to be extremely difficult to manufacture, and they had never been used in the extreme conditions of space. In the end it was decided that 2 of the 10 CCD chips would be backside-illuminated chips.

A few months later, in May of 1996, the ACIS team ran into more trouble. During testing at Lincoln Lab, one of the chips didn't work. Al Pillsbury, the engineer at Lincoln Lab who was responsible for the design and assembly of the CCD array, called Mark Bautz, Ricker's deputy, and a key member of the ACIS team.

"Things don't always turn out the way you plan them," Pillsbury said. "And I can prove it."

The problem was traced to flexprints, thin strips of metal that connect the CCDs to the electronic processors. The solder joints connecting the flexprints to the CCD chips had cracked during the rigorous thermal cycling tests that dunk the chips in liquid nitrogen until they cool down to −150 degrees Celsius (238 degrees below zero Fahrenheit), and heating them until they warm up to +60 degrees C (140 degrees F). Further testing showed that all the flexprints would have to be removed. The flexprints were attached with epoxy, and their removal posed formidable problems of contamination and destruction by electrostatic discharge.

"I thought it would be nearly impossible to replace them," Bautz said.

After weeks of agonizing over what to do, Pillsbury saved the day by coming up with a process for removing the flexprints by melting the epoxy with a hot knife. He and his team were able to do it, and save 12 of the 17 CCDs, more than enough to construct the CCD array. The problems didn't end there, though. One batch of flexprints after another proved to be flawed, as they got rejection rates up to 90 percent. Pillsbury came through again.

"Al rode herd on the flexprint vendors to assure that we got the parts we needed," Bautz said. "He is one of the unsung heroes of the project."

In May of 1997, ACIS arrived at long last at the X-ray Calibration Facility in time for a rushed but intensive 9 days of testing before it was shipped off to Boulder, Colorado. There it would be inserted, along with the HRC, into the science instrument module built by Ball Aerospace and Technology. Then, if the schedule held, all the components would come together at TRW in Redondo Beach, where the completed observatory would finally take shape in preparation for a launch in August of 1998.

22

An Observatory and a Name

TRW IS ONE of a handful of companies in the world that can justifiably call one of its facilities Space Park. TRW has been building spacecraft since before NASA was born. TRW built Pioneer I, which was launched a little more than 3 months after NASA was formed in 1958. Since then TRW has built more than 190 communications, defense, and scientific spacecraft, including the Einstein X-ray Observatory and the Compton Gamma Ray Observatory. The company has produced, integrated, and tested more than 130 communications payloads and subsystems, and integrated more than 550 experiments into their host spacecraft.

When TRW was selected as the prime contractor for AXAF, the groups at Harvard-Smithsonian and the Marshall Space Flight Center were pleased. They would be working with the same organization and with many of the same people with whom they had worked on the Einstein Observatory. As we followed the AXAF program from one subcontractor to another, it was clear that they too shared the scientists' confidence in TRW, from its systems management techniques to its engineering, scientific, and computer support personnel.

When we visited the TRW Space and Electronics Group's 200-acre Space Park facility in Redondo Beach, California, with its impressive assemblage of office buildings, laboratories, manufacturing, and test facilities in late October of 1997, the launch of AXAF was scheduled for less than a year away. The game was in the fourth quarter and the ball was in its hands. Like the franchise player that it was, that's where TRW wanted it. The spacecraft—a precedent-setting model of lightweight graphite construction—had been built, and had been bolted to the telescope. The solar panels built by Fokker in The Netherlands were at the facility, and TRW-built photovoltaic cells were being installed. A new type of bi-propellant rocket engine had been built especially for AXAF to lift it to its final orbit once it was deployed.

In a few months the instruments would arrive in the science instrument module from Ball Aerospace, and the all-important acoustic and thermal vacuum tests of the observatory as a whole would begin. The TRW team members would shake it and bake it, as Joe Payne, TRW's lead spacecraft engineer for AXAF put it. The large acoustic test facility, "with woofers that would be the envy of every teenager in America," according to Brooks McKinney, the TRW AXAF public relations officer, would expose the observatory to sounds and vibrations equivalent to what it would experience at launch. Then, in the thermal vacuum chamber, they would subject it to the temperature and density extremes of space, taking it down to 240 degrees Fahrenheit below zero and up to plus 140 degrees F.

"I think it [the testing] is going to be a piece of cake," Payne told us, then added, "but each problem at this point is more severe because of time. It is a very intense job of coordination. Everyone needs to understand their role, to look ahead, not just to tomorrow, but months ahead. You can't relax—for want of a bolt you can't proceed."

To keep on schedule, the TRW team was working two 10-hour shifts 6 days a week. "I've never been so proud in my life," Payne said in describing the work of his group. "You don't have to ask them. They have such ownership of their hardware."

Jim Korka reflected the sentiment of many who worked on AXAF. "I feel proud and possessive. I almost hate to let it go."

Steve Loer agreed. "My wife keeps saying to me, 'It's not like it's one of your children,' but it almost feels that way sometimes."

They all realized that this "child" had to leave the nest and fly. Getting the "bird," as they often called the spacecraft, out of the nest in a timely way was uppermost in the minds of the managers.

"It's going to be a challenge to complete the electrical tests of the bird in time," said Ralph Schilling. "One thing we know for sure is that there will be surprises downstream . . . Managing the schedule is one of the

Assembling the spacecraft bus, which includes the electrical power, propulsion, control, and data-handling subsystems for the spacecraft. (TRW.)

most important managing skills. You have to manage your work force, but also the availability of the facilities to test and assemble the many parts that have to come together."

Ed Wheeler, the program manager for AXAF at TRW, had worked on the Einstein Observatory and many other spacecraft during his 34 years at TRW. He was well aware of the demanding schedule they faced.

"We have to stay within budget, and on schedule," he said. "It's not going to be easy."

Wheeler was most concerned about integrating the hardware—putting it all together and making it work.

"There are 9,000 wires to put in and 22,000 connections to make, and thousands of lines of computer code to write."

In October of 1997, the schedule required that the spacecraft leave TRW in 8 months in order to make the proposed launch date of August 28, 1998. Were they on schedule?

"Sure," Wheeler replied.

A few weeks later this confidence had evaporated. The development of the software to test the spacecraft fell far behind schedule. This delay spilled over into problems with the availability of certain test facilities. Despite adding more programmers and other personnel, and going to three 8-hour shifts 7 days a week, they had to acknowledge that they were not going to meet the June 1, 1998, deadline.

NASA issued a terse announcement that "TRW Space and Electronics Group, Redondo Beach, CA, has notified NASA that it will be unable to deliver the Advanced X-ray Astrophysics Facility (AXAF) to NASA's Kennedy Space Center, FL, on June 1, 1998, as required by contract, because it has experienced delays in assembly and testing of the facility." The launch date, which had held steady for more than 5 years, slipped 3 and 1/2 months to December 3, 1998.

"The delay in delivery of the observatory is unfortunate," Fred Wojtalik, the AXAF project manager at the Marshall Space Flight Center said in a prepared statement that went on: "However, our first priority is to launch a world-class observatory which has been thoroughly tested and meets all requirements. We will work closely with TRW to ensure that happens."

Shortly thereafter TRW made a managerial change. Craig Staresinich, who had years of experience at managing programs in the phase just before launch, succeeded Wheeler as TRW's AXAF program manager.

"I tried to bring the disparate groups together so they could function as a team," Staresinich said. "We called a series of meetings and asked them what they thought we could do. They said, 'No 24-hour schedule with three 8-hour shifts. There are not three of us to go around. We would do much better to work two 10-hour shifts.'"

Staresinich tried to implement the change, and found that it was against company policy. "I took it to the top [Tim Hanneman, executive vice president in charge of TRW's Space and Electronics Group] and got that change instituted."

When we visited TRW again in April of 1998, we did so in an official capacity. In 1991 a contract had been awarded by NASA's Marshall Space

Chandra in the thermal vacuum test facility at TRW. (TRW.)

Flight Center to the Smithsonian Astrophysical Observatory to manage the science operations of AXAF. In 1996, this contract was modified to authorize the group to manage the operation of the satellite as well. This was the first time that a group outside of a NASA center had been given responsibility for operating a major mission. The Smithsonian AXAF team, which included key personnel from MIT and TRW, was headed by Harvey Tananbaum, with Claude Canizares as associate director. They were responsible for planning the observations, operating the observatory, processing the data received from the observatory, and providing technical and scientific support to scientific users. Together with Kathy Lestition and Kim Kowal we had, as part of the Harvard-Smithsonian team, formed a public education and outreach program.

The mood at TRW was upbeat. All the parts of the spacecraft had been assembled, from the solar panels to the science instrument module on the end. The acoustic test designed to simulate the sound pressure environment inside the Space Shuttle cargo bay during launch was a success. The astronaut crew of Jeff Ashby, Cady Coleman, Steve Hawley, Michel Tognini, and their commander, Eileen Collins, the first female commander of a Space Shuttle mission, were at Space Park for a media event to help celebrate the completion of the assembly.

"It is beautiful," Collins said, referring to the 45-feet-tall foil-covered observatory. "We'll make sure it gets safely into space, where it belongs."

"We've got Mr. Murphy back in the bottle. Let's hope we can keep him there," Wojtalik said, referring to the eponymous author of Murphy's Law.

But bottling up Mr. Murphy proved to be difficult. More software delays cropped up, related this time to the testing of the observatory's safe mode, in which the telescope does not point toward the sun, while the solar panels do point toward the sun. More than one spacecraft has failed because it did not go into safe mode when it encountered a problem, so simulations of failures to check that the spacecraft can safely go into and get out of safe mode were an essential part of prelaunch testing. Software is used to simulate a failure, for example a gyro failure, and then the onboard software is supposed to assess the problem and decide whether to go into safe mode, and if so when to come out. The software needed to perform these tests turned out to be far more complicated than expected, so crucial tests could not be performed on schedule.

This problem was soon dwarfed by another. On June 18, a battery of thermal vacuum tests had almost been completed without incident and was winding down. The last test was to open the ACIS door and check for light leaks in the spacecraft. Such a check is important to evaluate the effects of solar illumination when the spectrometer is in orbit. The door is opened through the action of heated wax in an enclosed tube with a piston. When the wax is heated, it expands and pushes on the piston, which turns a shaft that opens the door. This wax actuator, which sounds like something Rube Goldberg might have invented, has been used successfully for years in spacecraft. This time, however, the door stuck and a fail-safe disk in the actuator failed.

The ACIS camera was removed from the spacecraft and shipped back to Lockheed Martin Aerospace, where it had been designed and built. When the engineers tested it, the worst-case scenario was realized: the door opened.

"Now," Gerry Austin, the veteran space engineer explained, "they don't know what the problem was! They'll have to guess."

Although they checked it thoroughly for 5 months, the Lockheed Martin Aerospace engineers could never determine why the door didn't open in the test. The favorite theory was that the O-ring seal on the door stuck, possibly because the door was cold (−115°C or −175°F) and had not been opened for more than 7 months, but neither this nor any other theory could be proven.

"They ruled out every possibility," Tananbaum said. "You would have concluded from the tests that the door must have opened!"

The mechanism that opens the door was rebuilt and modified to give a greater margin of error.

"We expect the ACIS door to open on orbit, based on all our tests," Garmire said. "But then, the door had always opened before."

The effect of the software and the door problems was that the launch slipped again, to January 23, 1999.

Although TRW and the ACIS team were grappling with significant problems, NASA felt sufficiently confident that the launch date would hold. The time had come to give AXAF a new name commensurate with its anticipated greatness, just as NASA had named the Hubble Space Telescope after the famous astronomer Edwin Hubble. In keeping with the openness of the AXAF program from the beginning, it was decided that the choice of the name would be opened up to the world.

The Harvard-Smithsonian AXAF team was given the job of running the name contest, and one of us (KT) became the contest coordinator. TRW put up the grand prize of an all-expenses-paid trip to the launch. The contest was publicized in newspapers, at meetings, and on the World Wide Web. It drew more than 6,000 entries from 61 countries. Each entry was accompanied by a brief essay in support of its choice. An elite panel composed of three scientists, an aerospace official, two members of the science media corps, and a former secretary of defense selected the winning name and two winning essays, one by Tyrel Johnson, a high school student from Idaho, the other by Jatila van der Veen, a high school teacher from California.

Their choice for the observatory's name was Chandra, in honor of Subrahmanyan Chandrasekhar. Known to the world as Chandra, which means "moon" or "luminous one" in Sanskrit, Chandrasekhar was widely regarded as one of the foremost astrophysicists of the twentieth century. Chandra immigrated in 1937 from India to the United States, where he joined the faculty of the University of Chicago, on which he remained until his death in 1995. Chandra and his wife, Lalitha, became American citizens in 1953.

Early in his career, Chandra demonstrated that there is an upper limit—now called the Chandrasekhar limit—to the mass of a white dwarf star. A white dwarf, the last stage in the evolution of a star such as the sun, is formed when a star collapses after its nuclear energy resources are exhausted. This discovery is of fundamental importance to X-ray astronomy, since it shows that stars much more massive than the sun must collapse beyond the white dwarf stage to form either neutron stars or black holes in an explosive cataclysm. Exploding stars, neutron stars, and black holes are responsible for many of the strong X-ray sources in our galaxy and beyond.

Chandra's accomplishments were not confined to one field. He made important contributions to nearly all branches of theoretical astrophysics, and published 10 books, each covering a different topic, including one on the relationship between art and science. For 19 years he also found time to serve as editor of the *Astrophysical Journal* and transform it into the world's leading publication in the field of astrophysics. To hear him lec-

Subrahmanyan Chandrasekhar. (University of Chicago.)

ture in his prime, as one of us (WT) did at a summer school session, was not so much inspirational as awesome. To read his comments on a paper one had submitted to the *Astrophysical Journal* could be another matter altogether. In 1983, Chandra was awarded the Nobel Prize in physics, to no one's surprise.

The choice of name was greeted with widespread approval in the scientific community and beyond. The formal announcement of AXAF's new name was scheduled for September 1, 1998. But the announcement was delayed as NASA fretted over possible negative fallout as rumors persisted of more problems with the spacecraft. The safe mode testing still was not complete, electrical switches designed to switch power on and off between various components of the spacecraft were acting up, and an interface unit that handles communications between the onboard computers and other parts of the spacecraft failed to check out.

Tananbaum expressed concern, but in keeping with his perennial optimism, told us in late September, "They [the people at TRW] are making good progress. They should be able to hold the launch date. They could fix these problems at the Cape [Canaveral] if they had to," he said.

Edward Weiler, acting Associate Administrator for Space Science at NASA, decided not to ship. "Delaying shipment might mean spending an extra 10 to 20 million dollars," he said, "but the alternative was taking a risk with a 2-billion-dollar taxpayer investment. So I decided this was not the time to ship."

On October 13, 1998, NASA announced another delay, "to allow additional testing" and to replace one of the electrical switching boxes. NASA, concerned that three launch delays in 12 months might indicate serious systemic flaws, did what it always does in such situations—it formed a committee. NASA chief engineer Dr. Daniel Mulville would chair the committee, and report the results by mid-January 1999. Wojtalik repeated his statement that "our priority remains the safe and successful launch of a world-class observatory, which has been thoroughly tested and meets all requirements."

In late October, Tananbaum traveled to Washington to meet with NASA Administrator Dan Goldin and others to discuss the engineering options and weigh them against the potential scientific impact of the launch delays. He returned with a positive assessment that good progress was being made and the committee would have its report out by early November. Two days later he received more bad news. Two electrical relays of the type used on AXAF had been found to fail in tests on another spacecraft. These relays were critical—they would be used to open a door covering the telescope. To make matters worse, they were in a difficult-to-reach place, so their replacement could lead to further delays.

"I'm through telling people that we've solved all the major problems," he said one day over lunch at the Wok and Roll Chinese restaurant, one of his favorite dining spots near the observatory. "It's beginning to wear me down."

The next week brought reports of another electrical interface problem and more possible problems with relays. Then, things began to turn around. The problems with the relays were fixed, the faulty electrical interface units were replaced, and the software had been debugged. On December 21, 1998, NASA announced in a press release that AXAF would henceforth be known as the Chandra X-ray Observatory. The release also stated in somewhat convoluted, but as it turned out, prudent language that "the telescope is scheduled to be launched no earlier than April 8, 1999."

On January 14, we were at TRW's Space Park again, this time to participate in a "roll-out" ceremony to give the press a chance to see the completed, tested spacecraft before it was "bagged" in its protective cover and shipped to Cape Canaveral. After the ceremony, we talked to Joanne McGuire, the vice president of the division in charge of Chandra. We mentioned the beauty of the observatory, towering above us like a four-story piece of monumental art, and how relieved we were that it was finally heading for Cape Canaveral.

"Yes," she replied, "but we still have some issues with some circuit boards."

Translation: Another launch delay was coming. Weiler, who was now the official Associate Administrator for Space Science at NASA, once again decided that the problems should be fixed at TRW and not at the Cape.

"It's not the right environment to do this kind of critical testing," he explained.

This delay was considered sufficiently serious that NASA held a telephone press conference with key members of the "space press" at which it announced a launch slip of 5 weeks. Printed circuit boards of a type used on Chandra's command and data management system failed during testing on another spacecraft. Flaws in the manufacturing process had left a number of boards with bad connections. The failed boards and those in Chandra were all made about the same time by Gulton Data Systems of Albuquerque, a company that was subsequently purchased by B. F.

Inspecting Chandra in preparation for its shipment to the Kennedy Space Center. (TRW.)

Goodrich. Engineers examined Chandra's circuit diagrams and came back with the alarming report that Chandra had 129 of the suspect circuit boards.

Fortunately, only four of the circuit boards on Chandra required replacement, and they were in a relatively accessible location on the spacecraft. It was determined that they could be installed at Cape Canaveral after Chandra arrived there. On February 4, 1999, two U.S. Air Force C-5 Galaxy transport planes carried the observatory and its ground support equipment to the Kennedy Space Center on Cape Canaveral.

As the crew members prepared to unload Chandra, they were told to stop. While the C-5 transport had been flying from Los Angeles to the Cape, tests of the engineering model for the ACIS detector were underway at Lockheed Martin Aerospace. The engineers there were still concerned about whether the door would open, and were testing it under a variety of conditions. On one of the tests, the door had failed to open. When it was cooled down and heated up again, it opened. But would the flight model, which was now packaged securely inside the spacecraft, open or not when it was in space? The airplane sat on the tarmac in Florida for 2 days while the Chandra team weighed and debated their options. To send the observatory back to TRW and test the flight model would trigger another major delay. Flying without testing entailed a significant risk. The Chandra team members decided to test the engineering model one more time. If it opened, they would assume that they knew how to fix the problem of a sticky door—by cooling and heating it—and they would proceed. If not, they would turn the airplane around and fly back to TRW.

Nervous groups of scientists and engineers huddled around speaker phones at Marshall, TRW, NASA Headquarters, and Harvard-Smithsonian as they awaited word of the test from Lockheed Martin Aerospace.

"It opened," the Lockheed Martin engineer Neil Tice reported in a matter-of-fact voice. His announcement was followed by spontaneous cheers and audible sighs of relief.

In the course of the telephone press conference held 2 weeks earlier to announce the fourth launch delay, one of the reporters asked NASA official Kenneth Ledbetter whether Chandra's problems were due to lack of quality control, or bad luck, or the complexity of the observatory.

"It says something about the complexity," Ledbetter replied. "It is a complicated vehicle."

In February, we asked Tananbaum essentially the same question: Is Chandra simply too complex?

"The mirror was difficult to produce," he said, "but in itself it is not complex to operate. However, it was bonded and assembled in a complex way in 1 g (Earth's gravity) and must work in 0 g [a weightless state] after blastoff and thermal cycling."

But the mirror had been ready for well over a year, and had not caused a slip in the schedule. A more serious problem, paradoxically, was that computers had become very good.

"Because of the increase in power of computers," Tananbaum explained, "we now have real-time systems, asking for the temperature every fraction of a second. It's not necessary. We don't need to know most of the environmental parameters that often. This causes information overload that drives the system harder than is necessary. The availability of computers entices engineers to make the systems more complex . . . It would definitely be an advantage to have a simpler system."

Another part of Chandra's complexity was its propulsion system. The Space Shuttle could only carry Chandra into space. To reach its operating orbit, two other additional rocket engines would be required. About an hour after Chandra was pushed away from the Space Shuttle, Boeing's Inertial Upper Stage rocket engine would fire twice and take the spacecraft to a temporary orbit, after which time the Inertial Upper Stage would separate from the spacecraft. Then Chandra's own propulsion system would fire five times over the next 10 days to put Chandra into its final orbit.

The Inertial Upper Stage was considered a workhorse rocket engine by NASA and the U.S. Air Force. Since 1982, it had delivered 18 payloads to high Earth orbits and interplanetary trajectories. It had failed only once. On April 9, 3 weeks before Chandra was to be mated with its Inertial Upper Stage, an Inertial Upper Stage attached to a Titan IVB rocket failed, and left a U.S. Air Force satellite in a useless orbit. On April 26, NASA announced that Chandra's launch date, which had slipped once more to July 9, would slip yet again by an undetermined amount of time, pending an investigation by the U.S. Air Force. If the rocket failure was due to a fundamental flaw in the Inertial Upper Stage engines that had developed because of inadequate quality control, Chandra's launch might be put off for 6 months or a year.

Gloom descended over the Chandra team like a long New England winter. Before, the team members had always felt that by working together, sharing each group's expertise, and working very hard, they could track down and solve all the problems they encountered. Now there was a problem that was out of their hands, and they were hostage to a secret U.S. Air Force investigation. The mood was not helped by the knowledge that such investigations were rarely completed in less than 6 months.

"Chandra—the telescope that never launches—it just sits there on the pad," Tananbaum said mordantly, at the conclusion of a depressing lunchtime conversation on the prospects for a timely launch. He talked about pains in his knees, arms, and neck. "Maybe it's all related [to Chandra]. Maybe I need to slow down." Then, "I don't want to speak in public about Chandra, because they will ask when it will launch, and I don't know when it will launch."

"Some of us have been working so long on this project that not launching is the normal state," Leon van Speybroeck added, in a vain attempt at humor.

The pessimism extended to the top of NASA. "We may have a real problem here," NASA Administrator Dan Goldin was quoted as saying, "But we will not launch until it's safe. And if we have to wait a year, we have to wait a year."

As they waited for the results of the Air Force investigation, they moved ever closer to the the setting off of a domino effect that would postpone the launch for a year no matter what the outcome of the investigation.

The first domino was a scheduled overhaul of the only available Space Shuttle launch pad on the Eastern Range of the Kennedy Space Center in late July and early August. If Chandra couldn't launch before July 20, according to this schedule, it would have to be put off until August 8.

The second domino was the hurricane season, which began officially on August 17, and extended into September. The third domino was the shuttle manifest, which was full through December for a repair mission for the Hubble Space Telescope, and two other shuttle missions for the International Space Station.

The fourth domino was the need for the Columbia Space Shuttle's 10-million-mile tune-up. After five flights, Columbia was already 7 months overdue for extensive upgrades and modifications. With the Inertial Upper Stage booster attached, Chandra was so large it would fit in only one shuttle cargo bay, that of the Space Shuttle Columbia. If the launch of Columbia slipped into the year 2000, NASA would probably ship the shuttle to Palmdale, California, for a 9-month overhaul.

The inescapable conclusion: If the launch of Chandra/Columbia slipped past August 8 for any reason, the observatory would not go into space until the summer of 2000. The working out of this chain of events within NASA probably helped Chandra. An unnecessary delay due to a protracted air force investigation would not just jeopardize Chandra by having it sit on the ground and risk damage to the mirrors and instruments by contamination. It would postpone a surefire public relations bonanza associated with the flight of Eileen Collins, the first female commander of a Space Shuttle mission, and coincidentally, a U.S. Air Force colonel. NASA applied pressure from several different directions, and a deal was struck. The air force would not release the results of its investigation early, but it would make essential information available so that NASA could proceed with preparations for launch, if the results were encouraging. They were. The problem was traced to a faulty cable connection on the failed Inertial Upper Stage booster. The cables and other potentially troublesome components on Chandra's booster were quickly checked and replaced, and on Monday, July 12, the launch was set for "no earlier than July 20."

The launch date held, so we prepared to leave for Cape Canaveral early Friday morning. A series of press-related events was scheduled to start on Sunday, and there were passes to get, displays to set up, and so on. On Thursday night we received a call from Kathy Lestition. She said that a rumor was rampant that they were going to postpone the launch yet again. A capacitor of the type used on Chandra had failed on another mission. We immediately called Tananbaum.

He told us that TRW had found cracks in ceramic capacitors of the type used in the pointing system and the solid-state recorder that stored the electronic data from the images. A check of the duplicate capacitors at

TRW showed that they, too, were cracked. The big question was: Are the cracks serious, or are they merely cosmetic?

"We are having a telecon [teleconference] at eleven o'clock tonight," he said. "I'll call you when it's over."

We rebooked our flight from nine to noon, and awaited his call, which came at 2:00 A.M.

"My guess is that the problem will go away. I'm guessing it is 80-20 for approval. We should know by eleven o'clock tomorrow morning. They will be testing all night at TRW."

Indeed, the testing at TRW had been going on since Thursday morning, when Staresinich got word of the problem from one of his engineers.

"I was told by Fred Ricker [vice president of the electronics and technology division] and Tim Hanneman that the number-one priority in the company was to resolve this problem. We had a crew of 20 or 30 people working all night."

All space projects keep spares of virtually every part on the ground for contingencies such as this. These spares, or "witness coupons," as they are sometimes called, are part of the same lot as the flight parts, or were made and processed in exactly the same way as the flight parts, so they should show the same flaws as the flight parts. TRW engineers checked sample capacitors for every program they had which was flying the problematic capacitors. They found that all of them had a hairline crack on the edge, and that none of them had failed. Their conclusion: the cracks were cosmetic. Now, they had to make a decision. Should they test more, or go ahead with the launch?

"It was like walking on a steep mountain ridge," Staresinich said. "The worst thing is to launch when you shouldn't. The next worst thing is not to launch when you should. TRW's recommendation was that we should fly."

Was it a hard decision?

"Not for me it wasn't," Staresinich said. "I had the full support of management, and we had everyone in the program—from Weiler on down—consulting with us from the beginning. It was one of the great strengths of the program, the openness, the willingness for everyone to work together to resolve the issues."

At the eleven o'clock telecon the next morning, Alan Bunner, the science director of high-energy astrophysics programs at NASA, announced that the situation was still under review. When we pressed him for a decision in the light of all the planned activities at Cape Canaveral, he said, "I plan to get on a plane tomorrow and go to launch."

We left immediately for the airport.

VI

LOOKING OVER GALILEO'S SHOULDER

23

Launch

WE WERE STAYING in Titusville, a sleepy Florida town that wakes up from time to time when people come by the thousands to watch the beginnings of the space age. Titusville is a prime location to watch a rocket or shuttle launch, if you can't get on the base. The road from Titusville to the Kennedy Space Center press site on Cape Canaveral goes through the Merritt Island National Wildlife Refuge, the northern half of the barrier island that includes the cape. The refuge gave us a sense of space of another kind—open space for sea turtles, manatees, bald eagles, egrets, ibises, and alligators. For a few miles as we drove among the palmettos and sea oats we could imagine what it must have been like back in pre-Columbian times, when the Seminoles and other Native American tribes harvested oysters and clams in these estuaries. Then a right turn onto State Route 3 and we were on the cutting edge of the future—the Kennedy Space Center. A sign with a large "2" at the guardhouse reminded us that there were 2 days until launch.

We checked in at the press site, and met Donna Wyatt and Natanya Ness from Harvard-Smithsonian, who were there to help us set up our booth at the TRW press hospitality tent. By 10:30 A.M., the temperature was in the low nineties, and the humidity was not far behind. At 4:00 P.M. NASA held a press briefing that attracted about a dozen reporters, who listened to NASA officials comment on the readiness of the Space Shuttle and on details of the prelaunch procedures. The astronauts had already shifted their sleep patterns to an 11:00 A.M. to 7:00 P.M. schedule. At the time of the briefing we were a little more than 32 hours away from launch. An air force meteorologist gave the weather report—30 percent chance that bad weather could postpone the flight. The reporters, mostly veterans

At Launch Pad 39-B, the Chandra X-ray Observatory sits inside the payload bay of Space Shuttle Columbia. (NASA.)

of many Space Shuttle flights, asked questions about the problems with the Inertial Upper Stage, the cracked capacitors, the weight of the payload—the heaviest ever carried by the shuttle—and the tight schedule for launch. We looked at each other in wonder. Maybe, after 22 years of planning and politicking, thousands of technical problems, large and small, Chandra was going to be launched! It was beginning to get exciting.

A NASA science briefing was held Monday morning, July 19, at 9:30 A.M. Lalitha Chandrasekhar, Chandra's widow, and the name contest winners, Jatila van Der Veen and Tyrel Johnson, were introduced to the press, along with Tananbaum, Weisskopf, and the Chandra principal investigators: Bert Brinkman, Claude Canizares, Gordon Garmire, Steve Murray, and Leon van Speybroeck. The meteorologist repeated his forecast of a 30 percent chance of bad weather. After the briefing, the scene shifted to the TRW tent, where TRW had arranged a ceremony to honor Mrs. Chandrasekhar and the contest winners. Mrs. Chandrasekhar reminisced about her life with Chandra. "Chandra loved to look at the full moon," she said. "'We are like two halves of the moon,' he told me. 'Together we make a full moon.'"

"Would Chandra be pleased to have the observatory named after him?" she was asked.

"I doubt it very much," she replied. "He was a very private person."

The remainder of the day was taken up with tours, receptions, banquets, and family gatherings as the excitement and the carnival atmosphere grew. NASA distributed a list of celebrities that would be attending, ranging from First Lady Hillary Clinton to the U.S. Women's soccer team to Judy Collins, who had composed a song in honor of Eileen Collins, to model and TV commercial heartthrob Fabio. A tragic backdrop to the events was provided by an ironic twist of fate. John F. Kennedy, Jr., had been invited to attend the launch and weekend ceremonies at the Kennedy Space Center to celebrate the thirtieth anniversary of the launch of Apollo 11 on its journey toward the moon. He declined so that he could attend his cousin's wedding on Martha's Vineyard. Kennedy's plane crashed into the ocean on Friday night, taking his life along with those of his wife and her sister.

At a joint TRW and Marshall Space Flight Center event at Cape Canaveral on Monday evening, Martin Weisskopf acknowledged

Space Shuttle Columbia on the launch pad. (NASA.)

Riccardo Giacconi, who was present, as a leader in X-ray astronomy whose vision led to the development of the Chandra X-ray Observatory.

"It was a little strange," Leon van Speybroeck said. "I think a lot of people there didn't realize who he was, or what he had done, which is amazing to me."

On Monday night at 9:21 P.M. the astronauts boarded Columbia, and by

10:36 P.M. the hatch was closed and latched. The press room, which had been nearly deserted all day, began to fill up with reporters from around the world. It was 2 hours until launch. A little after midnight, we left the press room to wander around outside and find a good place to watch the launch. The old hands wouldn't be leaving the press room until about 10 minutes to launch, but we were too excited to pretend that we were old hands.

The press bleachers, which hold about 500 people, were mostly empty. We were told that the cognoscenti did not view the launch from the bleachers. They walk down to the lagoon about a hundred yards away and watch from the banks of the Turn Basin. We checked it out, decided that the improvement in the view did not offset the increased mosquito population, and returned to the bleachers. As the countdown dropped below 1 minute, our pulse rates rose, and despite the sweltering heat, our arms were covered with goose bumps. At T−10, everyone joined in with the countdown. The hydrogen burnoff pre-igniters activated, and then there was silence. A spontaneous groan arose from the crowd. Maybe it was just a momentary hold. Then we saw Ed Weiler striding briskly back from the Turn Basin, and we knew that the launch had been scrubbed.

At the 2:30 A.M. press conference, it was standing room only as NASA officials explained what had happened. A sensor designed to detect dangerous amounts of hydrogen gas near the engines gave a normal reading at T−31 seconds; then a spike in the read-out at T−16 seconds indicated dangerously high levels. Ozzie Fish, the engineer at the controls, had a difficult decision to make. Was the spike real or spurious? On the one hand, another reading at T−8 seconds would tell him. On the other hand, he had to make a call at T−10 seconds if he wanted to abort before the main engines started. He pushed the abort button. The T−8 reading indicated that the hydrogen level was back to normal, but the deed had been done. The launch was aborted at T−7 seconds, less than half a second before the main engines were to ignite.

A review of the data convinced NASA engineers that the reading had been spurious, but everyone agreed that Fish had made the right decision. The bad news was that the draining and refilling of the external fuel tank and the replacement of the pre-ignition burners would require a 48-hour turnaround. The good news was that the signal to stop the launch had been given in time to abort the liftoff before the main engines ignited. If the main engines had ignited, the turnaround time would have been several weeks, and any glitches in the replacement of the engines could have delayed the launch for a year.

We got back to our motel around 4:00 A.M., in time for a few hours sleep before returning for the 9:00 A.M. briefing. At the briefing NASA officials confirmed that the launch had been rescheduled for 12:28 A.M. on Thursday, July 22, with a launch window of 46 minutes extending to 1:14 A.M. This launch window was fixed by the need to get Chandra into the proper orbit with respect to the sun. We spent the rest of the day with friends and family—spanning four generations—who had gathered from many locations in the United States to watch the launch.

At the Wednesday morning briefing, Don McMonagle, NASA's shuttle launch integration manager, announced that everything looked good for the launch scheduled shortly after midnight. What about the hydrogen sensors? They had been thoroughly checked out. "Tonight we are going to let the engines start," he said. The meteorologist announced that there was only a 10 percent chance of bad weather interfering with the reloading of the external tank with 500,000 gallons of liquid propellant, and a "zero percent chance of bad weather at launch time."

At midnight, with lightning flickering in the distance, we were back in the bleachers, looking expectantly at the shuttle, easily visible at a distance of 4 miles, gleaming in floodlights. At 12:10 A.M. a 10-minute hold for weather was announced, pushing the launch back to 12:38 A.M. We returned to the press room to check on the Doppler radar map that was continuously displayed on one of the monitors. A small cell of bad weather was visible about 5 miles to the south, but it was expected to move out of the way quickly. At 12:30 A.M. the commentator announced that cloud-to-cloud lightning had been detected within 8 miles, but that the cell was still expected to move out of the way before the launch window closed. We went back outside to watch the storm go away. We could see lightning flashes in the clouds, and they didn't seem to be getting dimmer. We returned to the press room to look at the weather monitor. The size of the storm had increased dramatically.

The air force meteorologist, who had predicted a zero percent chance of bad weather, told us that the storm needed to move only 2 miles and it was moving at 12 miles an hour, so in 10 minutes it should be out of the way. "Then we need to wait another 15 minutes after the readings go down," he explained, then added apologetically, "I guess you think the weatherman ought to be fired." No one disagreed, though all appreciated his candor.

The orbital dynamics group huddled and concluded that the launch window could be extended to 1:20 A.M. and perhaps even to 1:30 A.M. At 1:19 A.M., lightning struck the ground 8 miles away. At 1:20 A.M., a "weather scrub" was announced. NASA would try again tomorrow night.

Now things began to get a little dicey. Concern over contamination of the mirrors and science instruments was growing with each passing day the observatory sat on the launch pad, and the shuttle support crew was exhausted, as was practically everyone at the Space Center. If NASA didn't launch on the third try, there would be a 5-day wait while a Delta II rocket carrying four cellular telephone satellites for the Globalstar consortium was launched. And if truly bad weather set in, Chandra would lose its place in the queue and the prospect of a year-long delay would loom again. The feeling was widespread that the third time had to be a charm. A call from Tananbaum on Thursday afternoon confirmed the feeling. "We just had a long meeting with people from headquarters," he said. "The engineers took a closer look and decided we could live with a longer launch window, so they're going to extend the launch window from 45 minutes to 1 hour and 56 minutes. They want us outta here."

So shortly before midnight on July 22 we arrived at the press site once more. Few people were still there since most of the national press had left to cover other stories. The TRW media tent and many of the support personnel were gone. Many of the distinguished guests, including the First Lady and Giacconi, had left, along with many other people who had to get back to their normal lives. The sense of expectation that had been prevalent in the NASA press room for the two previous launch attempts had been replaced by an air of quiet determination. Meteorologists had cautiously predicted a 30 percent chance of bad weather, but the night was clear and stars twinkled overhead.

Watching on the monitor in the grandstands, we saw the gantry on the shuttle support tower retract at T− 2 minutes and 30 seconds and counting. Like a modern-day cargo cult, people milling around the stands

Liftoff of Space Shuttle Columbia. (NASA.)

walked expectantly toward the Turn Basin to see if the big bird would fly tonight.

As the countdown dropped below 20 seconds, even the preternaturally calm voice of NASA launch commentator Lisa Malone rose a register. "Ten, nine, eight . . ."—the hydrogen burnoff pre-igniters flare into action—"seven, six . . ."—the main engines fire and the rainbirds around the launch platform unleash a torrent of water onto the platform to prevent a potentially damaging reflection of the intense sound waves from the rocket engines—"five, four, THREE"—the heat from the main engines vaporizes much of the water on the platform, sending a cloud of water vapor billowing outward—"TWO, ONE, LIFTOFF. WE HAVE LIFTOFF OF COLUMBIA, REACHING NEW HEIGHTS FOR WOMEN AND X-RAY ASTRONOMY!"

Shortly after liftoff at 12:31 A.M. on July 23, Columbia shone like the midday sun. Its reflection blazed in the dark water of the Turn Basin, and then it streaked outward over the Atlantic and upward into space for its rendezvous with history. Twenty seconds later, a volcanic, 125-decibel roar rolled over the press site, shaking the bleachers, our rib cages, and our whole being. By this time Columbia was traveling 514 miles per hour. When the main engines shut off 8 and 1/2 minutes after launch its speed was almost 17,000 miles per hour. Watching the shuttle accelerate away from Earth, we found one thought inescapable: Astronauts are extremely brave people. Another thought: The public, who pay a great deal of attention to the astronauts and very little attention to the scientists and their instruments are, after all, right. Going into space, learning how to break free from the Earth, is the main point of the shuttle flights. Telescopes such as Chandra are magnificent products of human vision, ingenuity, industry, and the thirst for knowledge, but on the Space Shuttle they are just excuses to go into space.

"It's great to be back in zero g again." Commander Eileen Collins radioed down from 155 miles above the Earth. It was great to have her and the crew there, showing us how to work in space, preparing us for the day when our descendants will venture out to live there, connecting us to the cosmos from which we came.

And, of course, it was also great to finally have Chandra where it belonged.

24

Activation

SPACE SCIENCE and exploration are risky, more so than anyone involved cares to admit. Five seconds after Columbia's launch, a short circuit lasting half a second disabled computers controlling two of the three engines, but back-up computers kept the engines working. Also, the pressure of the hydrogen gas in the engine compartment blew a small plug that had been used to patch the venerable but aging shuttle. Either problem had the potential to abort the mission and endanger the lives of the crew, so NASA grounded the entire shuttle fleet after the STS-93 Columbia mission to check for similar problems, especially for short circuits. Fortunately, the consequences for the STS-93 Columbia crew and its cargo were minor, but the hydrogen leak did leave the shuttle 7 miles short of its planned altitude.

The lower altitude was widely reported in the press, but it was never a major concern for the scientists. The Inertial Upper Stage Rocket engine attached to Chandra and Chandra's own integral propulsion system had more than enough power to take the spacecraft to its desired orbit. The concern among the Chandra team was that the Inertial Upper Stage had failed its last time out, and that the integral propulsion system, though thoroughly tested on ground, had never been used in space.

About 2 hours after launch, the payload bay was opened, and Chandra was allowed to come to equilibrium with its new environment. After the astronauts checked out the computers, heaters, and communication links, Chandra was raised to an angle of 58 degrees above the bay. Then, under the watchful eyes of Cady Coleman, the mission specialist in charge of deployment, Chandra was gently pushed away from the shuttle.

On flight day 5, the five STS-93 astronauts pose for the traditional inflight crew portrait on Columbia's mid-deck. In front are astronauts Eileen M. Collins, mission commander, and Michel Tognini, mission specialist representing France's Centre National d'Etudes Spatiales (CNES). Behind them are (from the left) astronauts Steven A. Hawley, mission specialist; Jeffrey S. Ashby, pilot; and Catherine G. (Cady) Coleman, mission specialist. (NASA.)

"It's so beautiful I wished we could hang on to it for a while," Coleman said.

The shuttle moved a safe distance away from Chandra. An hour later, controllers at Onizuka Air Force Base in Sunnyvale, Calfironia, gave the command to fire the first stage of the Inertial Upper Stage booster. The rocket burned for just over 2 minutes and then coasted for about 2 minutes. Then came the second burn, the one that had been fatal to the Deep Space Network satellite. The rocket fired for 2 minutes, and then shut off. The monitors in the Action Room at the Smithsonian Observatory's Chandra X-ray Center (CXC) in Cambridge, Massachusetts, showed that the booster had performed exactly as designed. Tracking and radar-ranging data from NASA's Deep Space Network confirmed that Chandra was in a proper transitional orbit that would take it more than 30,000 miles

Chandra in space after deployment (the spacecraft's shadow blocks out the middle of the telescope, and the solar panels have not been deployed). (NASA.)

above the Earth before swinging back to its closest approach of under 200 miles. A huge sigh of relief swept through the assembled managers in the Chandra X-ray Center Action Room, but the celebration was muted. They still had a long ways to go. Over the next 10 days, Chandra's own internal propulsion system would have to fire five times to boost it to its final working orbit—an ellipse in which Chandra is at an altitude of about 6,000 miles at its closest approach (perigee) to Earth and about 90,000 miles at its most distant point (apogee).

With the completion of the firing of and then the jettisoning of the Inertial Upper Stage booster, Chandra was flying on its own, and was controlled by the Smithsonian's Chandra X-ray Center. Personnel from Harvard-Smithsonian, MIT, and an experienced flight operations team from TRW were now responsible for commanding the observatory and conducting the science operations.

For months prior to launch, the Chandra operations and control team under the management of Roger Brissenden of Harvard-Smithsonian had conducted simulations in preparation for this moment. The training had been so intense that friction among team members from different institutions mounted to the point where a "teamwork" facilitator was brought in for an off-site retreat to conduct team-building exercises. "It worked," reported Brissenden, an Aussie whose low-key but effective style is especially well suited to a pressure-cooker atmosphere. "I have no doubt that we will be able to operate the satellite."

The moment of truth had arrived. The first of five planned firings to take Chandra to its operating orbit was scheduled for Saturday night, July 24. The Action Room at the Chandra X-ray Center was filled with scientists and managers from NASA headquarters, the Marshall Space Flight Center, TRW, MIT, and Harvard-Smithsonian, as well as members of the flight operations team. The mood was expectant and tense. Everyone knew, but no one said, that this was the first time that this particular rocket system had been used on a spacecraft. All conversation was conducted in hushed tones so that no one would miss the word from the flight controller that the burn had started.

At 9:15 P.M. word came that the burn had started. The hydrazine-fueled rocket burn lasted for about 5 minutes. During the burn, everyone hovered over the consoles, poring over the data as they flashed onto the screens.

"Everything is nominal," Jean Olivier, deputy program manager from NASA's Marshall Space Flight Center, announced dryly at the end of the burn. Translation: It worked! Cheers and applause erupted in the Action Room. The rocket firing took Chandra to an orbit that is 750 miles above Earth at its closest approach and 45,000 miles at its most distant point. In this orbit, Chandra took a little over 24 hours to make one complete trip around the Earth. The next rocket firing was scheduled for about 24 hours later.

"Four burns to go and four doors to open," announced Max Rosenthal of the Marshall Space Flight Center, referring to the doors to two instruments and the two doors that protected the telescope's mirror assembly.

The integral propulsion system rocket engines fired for the second time a day later for 11 minutes and 13 seconds, while Chandra was high above the Earth over the Atlantic Ocean.

"Six minutes, guys," Olivier said as he counted off the minutes of the burn from the program manager's seat. "Doesn't seem like a minute could be this long."

Finally, Olivier announced that the burn was over, and that "everything went down normal."

This rocket firing increased Chandra's speed by about 260 miles per hour, and took the spacecraft to an orbit that was 2,148 miles above the Earth at its closest approach), while leaving it at approximately 45,000 miles from Earth at its most distant point. The next firing, scheduled for 6 days later, would be the longest of the sequence, and would take Chandra a distance of almost 90,000 miles from Earth, about a third of the distance to the moon. This burn was necessary to lift Chandra far above the Van Allen radiation belts. The belts are caused by charged particles trapped in the Earth's magnetic field and range in altitude from 250 miles to 40,000 miles. They produce a blizzard of background radiation, or static, that makes observations of distant cosmic sources extremely difficult.

In the meantime, the door to the High-Resolution Camera was successfully opened, and Eileen Collins made a flawless night landing of the space shuttle Columbia at the Kennedy Space Center to complete the first shuttle mission ever commanded by a woman.

July 31, another Saturday night, found the leaders of the Chandra team huddled around computer monitors in the Action Room watching "the big one," as Steve Murray called it. The importance of this burn was evident from the number of high-level officials present on a summer weekend night: The Marshall Space Flight Center was represented by Fred Wojtalik, Chandra program manager, Jean Olivier, and Martin Weisskopf, project scientist. TRW, the prime contractor, had its program manager, Craig Staresinich, and its program scientist, Ralph Schilling, there. Tananbaum, Brissenden, and Murray from Harvard-Smithsonian were there, as was Claude Canizares from MIT. All told, the people in the room to watch this critical even represented about 150 man-years of work on Chandra.

The burn started at 6:33 P.M. Eastern Daylight Time, when Chandra was 2,000 miles above the Indian Ocean. When notification of the beginning of the burn flashed across the monitors, a ripple of excitement muted by tension spread throughout the room. Everyone crowded around the

Fred Wojtalik (left) and Craig Stairsinich (right) are pleased with what they see on the computer consoles after the critical rocket burn. (NASA/MSFC/K. Stephens.)

monitors, spoke softly, and watched columns of ever changing numbers, each one of which carried vital information about the progress of the burn and the state of the spacecraft—burn temperature, pressure, momentum unloading, velocity, delta velocity, and spacecraft orientation.

By 6:54 P.M., it was over.

"Exactly on time," Brissenden reported. "Everything appears to be nominal."

The group in the Action Room relaxed visibly, but Wojtalik, burned too many times by unforeseen difficulties over the course of the program, refused to celebrate.

"There are still more hurdles to clear," he warned. "We won't know for sure how good this burn was until they track the new orbit over the next several hours."

As it turned out, the burn went well, but not perfectly. The apogee of the orbit was about 559 miles short of the predicted apogee. This was within the specifications, but troubling, since it indicated that the rockets did not perform exactly as had been predicted. There was some concern that one of the engines, which had been burning slightly hotter than anticipated in the previous burns, might be wearing out. The difference was small, but real. Was it significant? A decision had to be made: stick with the existing engines, which had worked well—but not precisely—or switch to the redundant set of engines that had never been fired.

Upon advice from the TRW team that built the engines, Wojtalik decided to switch to the redundant engines, and postpone the fourth burn for 2 days while the software for making the switch was tested. What had been expected to be a routine burn was now awaited with anxiety. On August 4, at noon, the Chandra officials huddled once more around the monitors in the Action Room and watched the colored indicators as the two redundant rocket engines fired for the first time. No red lights flashed to warn that the engines were getting too hot, and when they cut off at 12:41 P.M., there was a palpable sense of relief in the room.

"It's great to have that one out of the way," said Roger Brissenden, Chandra X-ray Center manager. "I didn't expect anything to go wrong, but you never know. We were using different engines with different flow rates and different thrusts."

Hardly anyone showed up for the fifth and final firing early Saturday morning on August 7, which put Chandra in its operating orbit, ranging from approximately 6,000 miles (9,700 kilometers) at perigee to approximately 86,500 miles (139,200 kilometers) at apogee. This was partly due to the hour (1:43 A.M.) but was mostly a reflection that this burn to "trim" the final orbit would be a routine one. After a careful review of the orbit, and a brief "wiggle" of the observatory to check for the effects of sloshing fuel, the flight operations team permanently shut down the propulsion system and sealed it off from the rest of the spacecraft.

On Sunday afternoon, the Action Room was crowded again with Chandra brass from Marshall, Harvard-Smithsonian, MIT, Penn State, and TRW for the feature event: an outer space bake-off. This wasn't something sponsored by Pillsbury, but the result was even more delicious to the Chandra team than a two-layer devil's food cake. The goal was to warm up the housing of the Advanced CCD Imaging Spectrometer (ACIS), and then open the door of this X-ray camera that would be used to make the

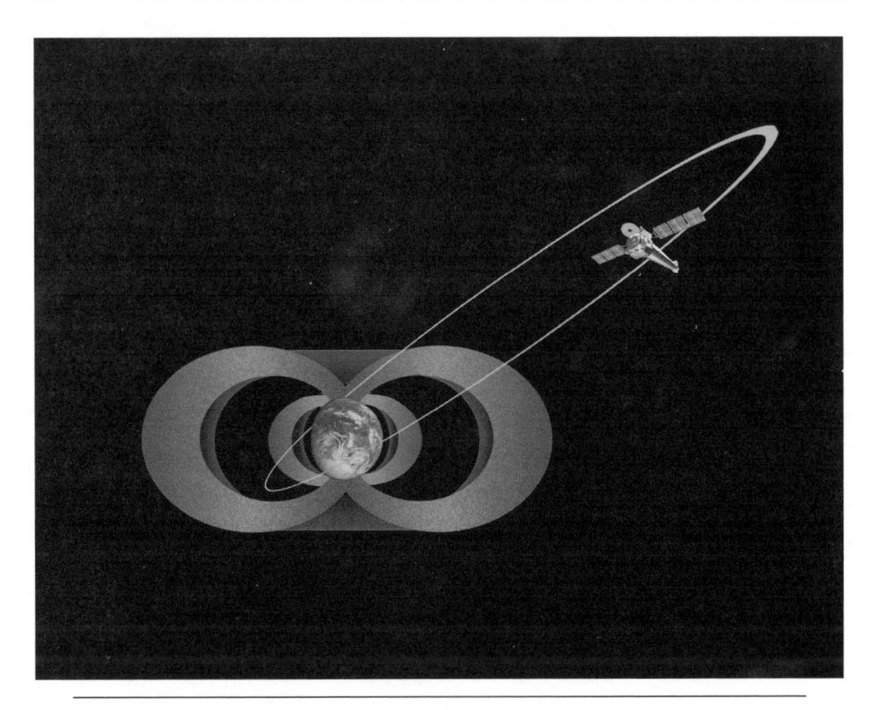

Schematic side view of Chandra's orbit, showing the inner and outer radiation belts. (SAO/CXC/M. Weiss.)

majority of the Chandra images during the observatory's first year of operation.

The plan was simple: Warm up the housing to evaporate any contaminants, then open the door to the camera. Yet one troubling historical fact hovered over the people in the room like an ominous summer afternoon thundercloud—this was the door that had failed to open during a prelaunch test and had caused a lengthy delay. Although the door mechanism was now repaired and tested, everyone was still nervous.

Before the operation even started, an argument about how to deal with a discrepancy in the procedure for opening the door broke out between a flight controller operating the observatory and the Chandra ACIS team. Every procedure for controlling the spacecraft has a detailed procedural flow chart and check list, and in this case there was a small discrepancy between the two lists. One of the flight controllers pointed this out, and

Roger Brissenden. (NASA/MSFC/K. Stephens.)

asked for permission to use what he considered to be the most up-to-date version of the procedure.

Gordon Garmire, the principal investigator for ACIS, objected. "We haven't practiced that procedure," he said.

The controller expressed concern about not following the updated procedure.

"Why didn't we have this conversation before now?" Weisskopf fretted in the Action Room.

The impasse hung in the air for about 15 seconds; then Brissenden spoke from his flight operations manager's seat.

"I concur with Gordon," he said. "Let's go with the way we have practiced."

"I'm more nervous than I was at the first IPS [integral propulsion system] burn," Martin Weisskopf said, as Paul Plucinsky from the Chandra Center announced that the commands to open the door were being sent to the spacecraft. At 5:12 P.M. the first heating cycle began, and the Action Room fell completely silent. The heater's job was to melt wax in a tiny piston on the ACIS door mechanism. The hydraulic pressure in the piston would then rotate a shaft that would open the door. One line on the figure-filled computer monitors would tell the story: the angle of rotation of the door shaft should show 70 degrees or more when it opened. Rather than risk a failure by opening the door on the first try, the managers planned to open the door in two or three separate cycles. They would first try to break any seal that might have formed in the extreme vacuum of space, and then they would open the door all the way. Each cycle would involve heating the wax to about room temperature and then letting it cool down again before starting again.

The first heating cycle produced no rotation. The second one, 16 minutes later, produced a 13-degree rotation.

"Needs to be at least 18 degrees to break the seal," Alan Bean, the Marshall engineer in charge of overseeing the work on Chandra's various doors and covers, announced anxiously. The third cycle produced a rotation of 19.5 degrees.

"Now, I know it's going to work," Weisskopf said, and Brissenden gave a thumbs up from the Control Room. Everyone agreed, but no one moved away from the monitors.

The fourth heat pulse opened the door to 36 degrees. Then at 6:46 P.M., on the fifth pulse, the shaft rotated to 71.5 degrees. Spontaneous clapping and cheering broke out. The door was open!

"Yes, I'm a little relieved," said Gordon Garmire, the Penn State University scientist who is the ACIS principal investigator, as we shook his sweaty palm. "Actually, I feel really good!"

Four days later, the rear contamination cover to the mirror assembly was successfully opened. After 20 years, the Chandra team was down to the last door, the sunshade door on the front of the telescope, which also served as a contamination cover.

On August 12, 1999, at 2:00 P.M. Eastern Daylight Time the Chandra X-ray Observatory received a command from the Chandra X-ray Center in

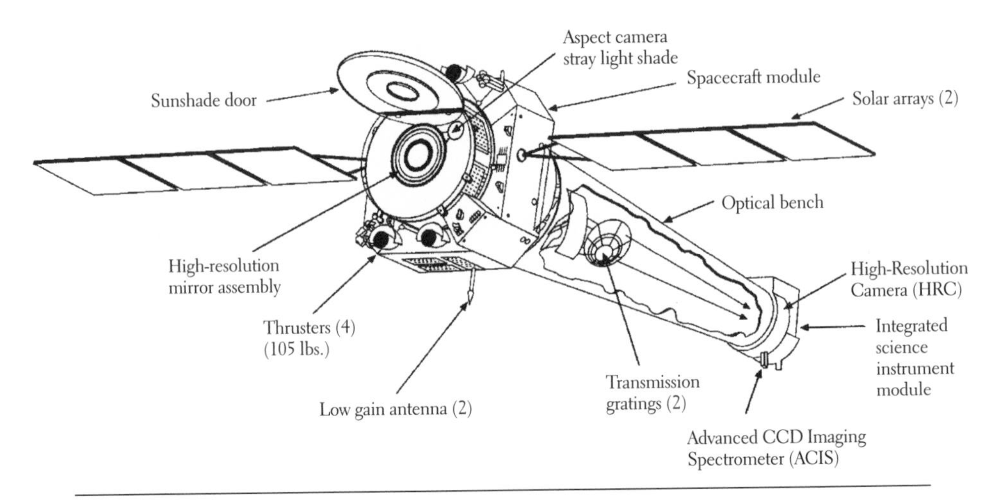

Schematic of Chandra showing various instruments and components. (SAO/CXC/TRW.)

Cambridge, Massachusetts, 84,000 miles away. A pyrotechnic charge equivalent to that of an M-80 firecracker exploded, sending a chisel through a bolt, and a powerful spring swung open the 120-pound, 9-feet-diameter door protecting the delicate mirrors of the observatory. Chandra wobbled slightly in its orbit, then settled down. Cosmic X-rays shone onto the mirrors for the first time.

"We still have a ways to go," Wojtalik said, and smiled as he realized that it was useless to make an official statement in the midst of the pandemonium that had broken out as it became clear that what had been a dream for two decades had become a reality. The flight controllers had to call for quiet several times, and they were soon abandoned as everyone left the Action Room for the ACIS instrument room, where the first X-ray photons would soon be visible on monitors.

BIG-TIME SPACE science has its drawbacks. Long, long waits, sometimes for decades before your project finally gets to fly. There are bruising political battles in which your instrument can get bumped off the bird. And there is always the ever present shadow of catastrophic failure, from which there is no recovery.

Why then do it? The answer was there for anyone to see at the Chandra

Artist's conception of Chandra in orbit. (TRW.)

X-ray Center on August 12. It was on dozens of faces that would not quit smiling. Faces of people that, during the previous 3 weeks, you would have sworn would break if you had asked them to smile as they endured two scrubbed launches, one successful launch that was almost aborted, two boosting burns with a type of rocket that had malfunctioned 4 months earlier, five more burns with a new propulsion system, and the remote opening of four separate doors—any one of whose failure to open could have terminated or seriously damaged the mission.

It was standing room only in the ACIS instrument room. Garmire was peering over the shoulder of his wife, Audrey, as she tapped on the keys of her computer and brought up a screen showing the response of ACIS.

"We are getting photons," Garmire announced breathlessly.

Photons from where?

"Mostly background," he cautioned, as a buzz started among the on-

Awaiting the first light from Chandra. From the left in the foreground: Mark Bautz, Beverly Lamarr, Gordon Garmire, and Audrey Garmire. In the background, facing forward: Jan Vrtilek. (SAO/CXC/K. Lestition.)

lookers. "But if the gyros hold us still long enough, we will probably see something."

Out in the technical support team, or TST, room, Staresinich reported, "We're controlling our gyros okay, but we still don't know where we are."

The aspect camera, another vital piece of equipment on Chandra, was recording the positions of stars and comparing them with the positions of 19 million stars, star clusters, and galaxies in the guide star catalog to find out exactly where Chandra was pointing.

"We have locked," Tom Aldcroft of Harvard-Smithsonian announced from his console.

"So have we!" Rob Cameron, also from Harvard-Smithsonian, yelled from another computer.

Staresinich looked over Cameron's shoulder and inquired, "This is good, isn't it?"

"This is better than good," Cameron replied.

"It is going so well that Chandra himself would have been pleased," Tananbaum observed, referring to the demanding standards Chandrasekhar set for himself.

"It is a great day for X-ray astronomy," said Dan Schwartz of Harvard-

Leon van Speybroeck. (NASA/MSFC/K. Stephens.)

Smithsonian, a 15-year veteran of the program and Chandra's science operations manager.

Back in the ACIS instrument room, George Chartas reported that they had found a source—67 photons so far—from somewhere out there in the direction of the south ecliptic pole.

"Audrey, have you named it yet?" asked Tananbaum.

About that time Leon van Speybroeck walked into the room. Van Speybroeck has been designing and overseeing the construction of X-ray mirrors longer than anyone on the planet and as the Chandra telescope scientist, he played a major role in the design and construction of Chandra's mirrors. He was there to see if the last 20 years of work on Chandra had been worth it, or if they were flying around with a "bucket of broken glass" as he put it.

The image on the screen was a little smeary.

"How big is that image?" someone asked.

"About 3 arc seconds, but it's off axis" (meaning that the source was slightly off to the side of where the telescope was looking).

"How big should it be, Leon?"

"How far off axis are we?" he asked.

"About 3 arc minutes."

"Then the source should have a size of about 3 arc seconds," van Speybroeck said with a broad grin. "We've got good mirrors."

"Come on, Leon, they're great mirrors," said Weisskopf. "And, that's the first source they've ever seen. Leon X-1!" he joked.

"Yeah, Leon X-1!" the group agreed.

Van Speybroeck was still smiling when we left the room about an hour later.

"It's a good start," he acknowledged. "Not a bad start at all."

25

First Light

AFTER THE SUNSHADE door opened to let the first cosmic X-rays shine into the Chandra Observatory, the members of the calibration team began to test out the observatory. They moved it around, looked at sources directly on the axis of the mirror assembly, then looked off axis. They shifted the Advanced CCD Imaging Spectrometer (ACIS) a few fractions of a millimeter at a time, searching for focus in much the same way that you can watch letters come into focus as you move a reading lens up and down over a page of printed words.

A quasar called PKS 0637–752 that is billions of light years away had looked like a point source within the limits of the resolving ability of previous X-ray telescopes, so it had been chosen as a source to test how smeared out the Chandra image would be off axis, or out of focus. To the surprise of the calibration team, the quasar split into two components as the team members moved the detector close to the expected focal surface. Plainly visible were the central quasar and a flamelike jet flaring out from the quasar (Plate VID). A quick calculation showed that the jet must be more than 200,000 light years in length, more than twice the diameter of our Milky Way galaxy. A check of observations of this object with a powerful radio telescope confirmed that it had also seen the jet in radio waves, so there was no doubt about its reality. There was also no doubt that they were very close to achieving focus.

The Chandra team prepared to point toward Cassiopeia A (Cas A), the remnant of a supernova that would be the target for the first significant image made at the focus. Cas A is a strong source of X-rays and has been observed by every previous X-ray telescope, sometimes for as long as 2 days. The Chandra image of Cas A would provide the Chandra team with an immediate and demanding test as to just how good their observatory was. Video crews were scheduled to cover the anticipated event on Tuesday, August 17. Then came a reminder that this was, after all, space science.

"We've had a major hiccup." It was Harvey Tananbaum calling to say that Chandra had gone into safe mode.

"A load (a series of commands radioed up to the spacecraft) put two incompatible maneuvers back to back," he continued. "The spacecraft didn't like it and went into safe mode."

In safe mode the instruments are powered down, the spacecraft points so that its solar panels get maximum sunlight, and the mirrors point away from the Earth or the sun. One of the problems encountered early in the Hubble Space Telescope's mission was the spacecraft's recurrent entry into the safe mode. This procedure protects the spacecraft and its instruments from getting too cold or too hot. The concern is that all the instruments have to be powered down and than back up again and computer instructions have to be reloaded, all of which increases the possibility of error or malfunction.

"Recovery will be done very carefully," Tananbaum emphasized. "It could take 6 hours. It could take 48."

As it turned out, it took around 30 hours. By Thursday, Chandra was out of safe mode. The mood in the Action Room was relaxed, the scientists were making small talk, and Tananbaum was checking in from time to time on a Red Sox game. Around 8:00 P.M., the command was made to slew toward Cassiopeia A. The small talk stopped and all eyes went to the monitor, looking for a sign that the aspect cameras would lock on guide stars.

"Three stars. That'll hold it!" Tananbaum exclaimed, as the "flags" on the monitor lit up.

"We're there! A full set of flags!" shouted Roger Brissenden.

The monitors were quickly abandoned as the scientists rushed out of the Action Room, into the technical support team room, past the rows of computer monitors with engineers and scientists staring intently at the screens, and into the small instrument room that had been the scene of so much excitement when the observatory had received its first X-rays.

"How long before we get an image?"

Harvey Tananbaum (left) and Stephen Murray. (SAO/CXC/K. Lestition.)

"About 40 minutes," replied Mark Bautz.

People dispersed for a while, moving over to the snack room to get a bite to eat, or into a nearby office to check on their e-mail or make calls with their cell phones. At 8:40 P.M. the room was full again. The X-ray photons were beginning to stream in at the rate of about 300 per second.

Then, there it was! A gorgeous, dramatic image of the remains of a star that exploded 10,000 light years away. Within half an hour we were looking at the best X-ray image ever made of a cosmic object (Plate I). Thirty times better. Comparable to the increase in resolving power of the Hubble Space Telescope over previous optical telescopes. Comparable to the increase in Galileo's telescope over the human eye. It was as if we were looking over Galileo's shoulder, peering through the eyepiece of his telescope, and getting a view of the universe that no one had ever seen before. Words cannot describe the feeling, but people tried.

"Awesome!"

"Spectacular!"

"What's that in the middle?"

What indeed? Close inspection showed a tiny bright dot in the middle of the remnant. Was it real? Yes. Was it the long-sought neutron star or black hole that was produced in the explosion that created Cas A? Possibly. Chandra will look at Cas A again for a much longer time as part of the scientific observing program. After those data have been analyzed, and scientists check for a telltale periodicity in the X-rays, as one might expect from a rotating neutron star, or a peculiar spectrum, or distribution of the X-rays with energy that would signal a black hole, or a neutron star, we should know the answer.

For now, the case had been made. In about an hour of observing, Chandra had produced an image incomparably better than that delivered by any previous X-ray telescope, and had discovered something at the center of Cas A that had never been seen before.

Over the next 2 weeks the High-Energy Transmission Grating Spectrometers built by Claude Canizares and his MIT team, and the Low-Energy Transmission Grating Spectrometer built by a team led by Peter Predehl of the Max Planck Insititute in Germany under the direction of Bert Brinkman of the Space Research Organization of the Netherlands were swung into place to look at Capella, 40 light years from Earth and the sixth brightest star in the northern sky. Actually, Capella is not one star, but two in orbit around each other. Both stars are normal stars that have extremely hot upper atmospheres with temperatures of several million degrees. The hot atmosphere, or corona, of Capella is of interest to X-ray astronomers because it is several times hotter and brighter in X-rays than the sun's corona, but in these first observations the two grating spectrometer teams were just checking out the capability of their instruments by focusing on a well-known source. They were not disappointed.

"Within the first hour we had obtained the best X-ray spectrum ever recorded for a celestial source," said Canizares.

The first look at the richly detailed spectrum returned by Chandra's gratings showed that what had been a jumbled forest of spectral lines was now a well-resolved stand of "trees" caused by the presence of silicon, magnesium, neon, and iron. It will take a while to sort it all out, but observations with the gratings will help to lay the foundation for a greatly improved theoretical understanding of the X-radiation from hot gases. Such an understanding is crucial for accurate interpretation of many of the observations Chandra and future X-ray observatories will make. The ob-

Claude Canizares. (NASA/MSFC/K. Stephens.)

served conditions in the cosmic sources are difficult if not impossible to reproduce in a terrestrial laboratory, so observations of Capella and other stars are in essence laboratory experiments in atomic physics.

On the same day that the high-energy grating was activated, the High-Resolution Camera was placed into the focus for the first time. It was a nervous moment for Steve Murray, the principal investigator for the HRC.

"I had a lot of anxiety," he said. "I was thinking of all the things that could go wrong. A bad pointing could put too bright of a source in focus and we could burn a hole in the camera. When the data came down, the first thing that went through my mind was, 'Oh my God . . . nothing's wrong!' It was such a relief, like being able to exhale after holding your breath for 23 years."

Gerry Austin, the project engineer acting for the High-Resolution Camera, had no such qualms.

"I never had any expectation other than it would work," he said. "You shouldn't have to be keeping your fingers crossed. I've had too much experience to believe that they're not going to work when they get up there. We worked too long and hard for any other result."

Still, Murray believed that the result could have been better.

"I kick myself for not doing things better," he said. "For example, we have an electronic timing problem, so we are not getting rid of the cosmic ray background as well as possible, and there is some electronic blurring that could've been fixed by adding one capacitor in the electronics . . . I build instruments for a living, and I want them to be perfect."

The High-Resolution Camera is not perfect—the increased cosmic ray background noise has a significant effect on the spectroscopy portion of the detector that can fortunately be mitigated by changing the software used to control the electronics on the detector. This incident served as a reminder that no instrument works in space exactly the way it did in the laboratory tests on Earth, and that the conditions under which instruments must operate are never exactly the same in space as on the ground—which is why they have to be recalibrated once they are in orbit. With the knowledge gained from in-orbit calibration, the problems mentioned by Murray have been sorted out, and the High-Resolution Camera has produced the sharpest X-ray images ever observed.

The responsibility for checking out the instruments in orbit rested on the Chandra calibration team. The cal team, as it was called, was a group of scientists that met daily for the first month after launch to assess the state and performance of the Chandra X-ray Observatory. Before launch, the performances of separate components of the observatory were thoroughly tested in various laboratories—the telescope at NASA's Marshall Space Flight Center in Huntsville, Alabama, and the spacecraft at the TRW Space and Electronics Group's Space Park in Redondo Beach, California. The cal team, under the direction of Christine Jones, had worked long hours during these prelaunch tests to understand the performance of the telescope under known conditions, in order to compare the performance of the mirrors and detectors in space with the test measurements.

Now that Chandra had launched, the cal team members moved on to their next phase—the 6-to-8–week in-orbit calibration phase, where they compared the performance of Chandra in space with what they expected from the testing on the ground, and with what previous observations of cosmic sources have shown. The expected and the real are never quite the same. Launch, the out-gassing of water vapor from epoxy or other materials, the hazardous trips through the radiation belts for 10 hours every 2 and 1/2 days, the bursts of high-energy particles released by solar flares, an unexpected bug in a computer program—all can cause problems. The process is something like driving a car in a simulator, and then getting behind the wheel of a car with bugs on the windshield, and driving on a road with potholes, shadows, rain, snow, and eighteen-wheelers.

The cal team's job was to sort all this out, literally on the fly. On one typical day during the first 6 weeks of operation, a dozen or more cal team members packed into the tea room at the Harvard-Smithsonian Observatory. Jones, the leader, started the meetings by saying hello to her colleagues who were patched in remotely from MIT and the Marshall Space Flight Center.

"This was a more or less quiet day except for the almost safe mode," she reported, referring to a minor glitch when the star tracker used to orient the observatory got confused.

Larry David commented on the numbers on the sensitivity of the detectors, and on the new results on the gratings. Brian McNamara commented on the point response function and wondered about the aspect—how well did they know where Chandra is looking? Herman Marshall of MIT broke in over the phone to mention a memo that addressed the situation. A graph was laid on the table and everyone huddled around to look at it and assess its meaning. On that day, the puzzle was quickly resolved. But on any given day, there were new challenges and new puzzles, some trivial, some not. Sometimes a change in the observing plans was required to test out a hypothesis.

"You never know what each day will bring," Jones commented. "Life is complicated, and so is Chandra."

On that very evening of September 8, 1999, life got a lot more complicated. McNamara, the ACIS specialist on the cal team, remembered it clearly.

"It was a horrible time," he said. "Paul Plucinsky called to say that the [ACIS] spectrum of Cas A didn't look right. I went over to the OCC [operations and control center] right away."

The problem was that the ability of ACIS to accurately measure the energies of the individual X-rays from the hot gas in Cas A had deteriorated, and was continuing to deteriorate. At first light, ACIS could measure the energy of an incoming X-ray to greater than 98 percent accuracy. Plucinsky's analysis indicated that this had slipped to about 90 percent.

Beverly Lamarr of MIT's ACIS team diagnosed the problem as being caused by a sudden increase in "charge transfer inefficiency."

A CCD works like a bucket brigade. An incoming X-ray creates an accumulation of charge at a specific location, called a pixel, on the silicon chip. The amount of charge accumulated increases with the energy of the X-ray. An applied voltage then moves the charge from one pixel to another, as in a bucket brigade, off the chip, where it is transformed into a signal. The timing of the signal tells how many times the "bucket" was handed off, which gives the location of the incident X-ray. The strength of the signal tells how much charge was handed off, which gives the energy of the X-ray. The process of moving the charge is not perfectly efficient—the inefficiency is analogous to a small, unknown amount of water being spilled with each hand-off in the bucket brigade—so the measurement of the initial accumulation of charge, and hence of the energy, is uncertain, depending on the charge transfer inefficiency. The ACIS team had seen the charge transfer inefficiency increase a thousand-fold.

"I was pretty sure it was due to radiation damage of some kind," Bautz said, "but I had no idea what had caused the damage."

A telecon was hastily convened at 9:00 P.M. with Wojtalik, Martin Weisskopf, Brissenden, Garmire, Bautz, Lamarr, McNamara, and others to discuss strategies for diagnosing possible causes and solutions. In the days that followed, it was quickly determined that all eight of the front-illuminated CCD chips were involved in the deterioration. The two back-illuminated chips were not. Since the back-illuminated chips have an extra protective layer that absorbs low-energy particles, this finding immediately pointed to low-energy charged particles, such as electrons or protons, as the culprits. These particles are abundant in the radiation belts through which Chandra passes for about 10 hours during each 62-hour orbit.

Other data seemed to argue against this interpretation. The particle fluxes were unexpectedly high, even for the radiation belts, and no damage occurred when ACIS was at the focus during belt passages. Rather, it happened when ACIS was shielded from the telescope.

A crisis atmosphere enveloped the Chandra Center. While it was true that most of the observations would use only one CCD chip at a time, the prospect of losing 8 of 10 chips on the X-ray camera that had been scheduled to make the majority of the observations was alarming.

"Gordon Garmire and I were both afraid that we might lose the entire instrument," Bautz recalled.

Telecons were held daily. Changing the orbit was not an option. It was decided that on September 13 the ACIS would raise the temperature of ACIS from -100 degrees Celsius, its normal operating temperature, to 30 degrees Celsius in an effort to "bake out" any irregularities in the structure of the chips. This procedure failed to improve the energy resolution.

"The news on the home front is not good," Tananbaum told us. He was concerned not only about the ultimate impact on Chandra, but about the political fallout. Chandra was being touted on Capitol Hill as a shining example of NASA's success, but the ACIS problems could make Chandra a political liability. As Tananbaum put it, "If this keeps up, we're going from the penthouse to the outhouse."

"I was devastated," Bautz said. "I was very busy in the beginning, and this probably helped me get through the first few days. I remember thinking at the time that if I were 20 years younger when this had happened, I might not have had the emotional wherewithal to deal with it."

As the days passed, the deterioration slowly continued, as the scientists and engineers frantically searched for an explanation. Desperation and despair began to set in.

"We had worked for 5 years to produce the best-calibrated X-ray astronomy instrument ever launched," Bautz said. "Now, before any astronomer could benefit from that effort, [it looked as though] much of our work was rendered irrelevant . . . After a particularly grueling multihour Sunday evening telecon, I can remember believing for the first time that the exquisite performance of our front-illuminated detectors was irretrievably lost. I wept."

Then, on Saturday afternoon, September 18, McNamara noticed that the chips had stopped degrading. He drafted an e-mail to the cal team and science operations team that said, "I find no significant change in the gain or CTI [charge transfer inefficiency]."

Bautz responded, "Best news I've had for a while," and asked for confirmation.

Catherine Grant of MIT answered within hours. "I noticed the same thing," she said.

The chips remained stable, but as McNamara told us, "It's small consolation, because we don't really know the reason—is it a fluke?"

Weisskopf insisted that the damaging radiation must be coming through the telescope, despite evidence to the contrary. With more data, Bautz and his team established that the degradation associated with the CTI was in fact related to passages through the radiation belts. The earlier data had been misinterpreted, partly because a transmission grating had been in the beam during some of the radiation belt passages and had mitigated the damage. Bronek Dichter, a U.S. Air Force expert on CCD chips, and part of a "tiger team" called in to help analyze the problem, suggested that the damage was due to the focusing of low-energy protons by Chandra's mirrors, which would explain the higher than expected particle fluxes. The magnetic brooms would remove the electrons, but not the heavier protons.

When ACIS was moved out of the focus during radiation belt passages, the degradation stopped. As of this writing, it has not recurred. Laboratory and on-orbit tests are under way to explore techniques for recovering some of the lost capability, which affects the precision with which the X-ray energies can be measured but not the quality of the images.

"There's no doubt at all in my mind that things don't always turn out the way you plan them," a greatly relieved Bautz said. "I can prove it."

Tananbaum was also realistic and well aware that other problems would arise.

"It's like having a baby," he said. "You have to stay up most of the night at first, then it gradually gets better, but that doesn't mean you quit worrying. You never quit worrying."

Meanwhile, Chandra's X-ray images continue to amaze and excite. The calibration team learns more about Chandra every week, and the mission planning team and science operations teams under the direction of experienced Chandra hands Bill Forman and Dan Schwartz make sure that the observations are carefully planned and efficiently executed by the flight operations team. Handling the flood of data from an observatory of Chandra's complexity is an enormous challenge, and a crucial link in the chain from observation to scientific understanding. The data systems group under the direction of Giuseppina Fabbiano and the science data systems team led by Martin Elvis are constantly updating the software to handle the data, to incorporate the latest information from the calibration team, and to implement improvements in the rapidly changing field of image processing. Thanks to the long hours put in by more than a hundred people in the Chandra X-ray Center, Chandra should produce magnificent images and make important discoveries for the next 5 to 10 years. The dream of a few X-ray astronomers has finally become a reality thanks to the effort, skill, and dedication of thousands of men and women in industry, universities, research institutions, and NASA.

Charlie Pellerin, the consummate NASA politician who was instrumental in getting and keeping funding for Chandra during critical periods, is now a management consultant. He still admires NASA's potential to inspire people.

"The genius of NASA is getting ordinary people organized in ways [so] that extraordinary things occur," he said.

Tananbaum, in addition to his duties as director of the Chandra X-ray Center, is already at work with a large consortium of X-ray astronomers planning and promoting the next generation of X-ray telescopes—a proposed fleet of observatories called Constellation X that as a group would have more collecting area than Chandra. They would not, however, match the resolving power of Chandra's mirrors. The judgment of the X-ray astronomy power brokers and opinion-makers was that it would be too expensive and hence not feasible politically to make another Chandra.

Van Speybroeck, while giving the Constellation X team the benefit of his unique experience in mirror design, is still not convinced that there should not be another Chandra. Perhaps because he has thought about and lived with the excruciating demands of building Chandra's mirrors, he seems more enthralled than others with the images they produce. He would stop by our office often to have a look at the latest colorized and enhanced cosmic portrait prepared by Dana Berry from the Chandra data. One day, the Chandra image on our monitor was of the Crab Nebula. Van Speybroeck, a man who loves science and the precision of equations, and who will tell you he has little use for philosophy and poetry, stood silently and took in the beauty and the wonder.

"That's a beautiful sight," he said at length, quietly, almost reverently. "We really have to build another one."

26

The Exploration Begins

THE FIRST IMAGES made by the Chandra X-ray Observatory were calibration targets. These targets, of which there were about 30, were mostly well-known cosmic X-ray sources. Some were observed many times and ways during the checkout period of the observatory and the data were made available to all interested parties.

After the calibration period ended, Chandra observing time was divided between a group of scientists with guaranteed time—principal investigators for the X-ray cameras and gratings, the telescope scientist, and six interdisciplinary scientists chosen by peer review—and scientists whose proposals are chosen each year by a peer review process conducted by a group under the direction of Fred Seward and Belinda Wilkes. These data remain the property of the scientists for 1 year, after which time the information is put in a public archive.

The early results from Chandra demonstrate the proof of the concept that was sold to NASA and Congress two decades ago. The images are stunning and the spectra are rich with detailed information about the temperature, density, and abundances of the elements in hot gas around normal stars, exploded stars, and black holes. In some cases, this information has already led to a clearer understanding and a confirmation of theoretical ideas; in others the data have raised new problems with the theories that may take years to resolve. What is clear is that information from Chandra will be a vital component of our search to answer some of the fundamental questions about our universe that Charlie Pellerin's group of experts posed 15 years ago in the Great Observatories booklet: questions about the life cycles of stars, black holes, and quasars, and about the formation and evolution of galaxies and galaxy clusters.

One of Chandra's early targets was the brilliant Orion star cluster, a cosmic birthing ground for stars. Stars in the Orion cluster were formed during the past few million years, so they are mere infants compared to our 4.5-billion-year-old sun. In one observation of the Trapezium region of the cluster using the Advanced CCD Imaging Spectrometer (ACIS) X-ray camera, Chandra detected a thousand X-ray stars. Some are well known, massive, optically bright stars, but over a dozen of the Orion X-ray sources have no known counterparts, even though sensitive optical and infrared telescopes have looked for them. These sources may be young brown dwarf stars. Such stars have masses so small (roughly 1/20th of the sun's mass) that they cannot sustain the nuclear reactions in their interiors and become true stars.

Most of the stars detected are young stars with masses similar to, or somewhat larger than, the mass of the sun. They give us insight into the way the sun looked a million or so years after it formed. Young stars, such as those found in Orion, are known to be much brighter in X-rays than middle-aged stars such as the sun. The elevated X-ray emission is thought to arise from violent flares in strong magnetic fields near the surfaces of young stars. The sun itself was probably thousands of times brighter in X-rays during its first few million years. Although the enhanced magnetic activity of young stars has been known for some time, working out the cause in any detail is a complex problem requiring much more knowledge about the behavior of turbulent gases in a magnetic field. With hundreds of stars observed simultaneously and representing a wide range of properties such as mass and rotation rates, the Orion nebula provides a valuable cosmic "experiment" for testing theories and computer simulations.

Eric Feigelson, a member of the Penn State ACIS team and leader of a group of scientists analyzing the Chandra observation of Orion, emphasized that the Chandra results could tell us more about young stars than just their flaring activity.

"X-ray astronomy now penetrates as deeply into the clouds as the best infrared and optical telescopes, permitting us to study high-energy processes during the earliest phases of star formation," he said.

Indeed, some of the Orion sources are still embedded in the cloud of dust and gas from which they formed; they can be seen with an infrared telescope, but not an optical one. Chandra shows that they are associated

Chandra image of the Orion star cluster. (NASA/Pennsylvania State University.)

with a distinct cluster of higher-mass stars deeply embedded within the murky Orion Molecular Cloud, marking the first time X-ray observations have been able to pick out individual massive stars still embedded in the clouds from which they were formed. More detailed studies of these objects may be important for understanding how planetary systems form.

"Recent studies suggest that when X-rays from young stars illuminate the disk around them, they may transform the properties of the disk and influence the formation of planetary systems," Feigelson explained.

If the early stages of a star's development are turbulent and stormy, the final stages can be full of sound and fury, and full of significance for our own existence. Carbon, nitrogen, oxygen, and other heavy elements necessary for life are created in the interior of massive stars. When a massive star runs out of fuel, it undergoes a catastrophic explosion called a supernova.

The matter blasted into space by a supernova creates a shell of multimillion-degree gas called a supernova remnant. This hot gas will expand and produce X-radiation for thousands of years. Chandra is the ideal instrument for studying these remnants.

Cassiopeia A (Cas A) is a prime example. It is the 300-year-old remnant of the explosion of a massive star. (Since Cas A is about 10,000 light years from Earth, we see events there as they occurred 10,000 years ago; the explosion that produced Cas A actually occurred about 10,300 years ago.) Previous X-ray images have shown an expanding shell that is about 10 light years in diameter, and has a temperature of about 50 million degrees Celsius. The material from the explosion is rushing outward at supersonic speeds in excess of 20 million kilometers per hour (12 million miles per hour). As this matter crashes into gas that surrounded the former star, shock waves analogous to awesome sonic booms heat the gas and the ejected matter.

The Chandra image of Cas A (Plate I), made with the Advanced CCD Imaging Spectrometer (ACIS), shows two shock waves: a fast outer shock, shown in red, and a slower inner shock, marked by the yellow areas. The brightest areas are the regions where the intensity of X-rays is greatest. Astronomers think that the star that exploded to form Cas A went through an unstable period a million or so years before the explosion, during which time its outer layers evaporated in a stellar wind. This stellar wind formed a bubble around the star. The inner shock wave visible in the Chandra image is believed to be due to the collision of the material ejected in the supernova explosion with this bubble.

The explosion that produced Cas A has proven to be an enigma. Although radio, optical, and X-ray observations of the remnant indicate that it was a powerful event, the initial outburst was not widely observed, if it was observed at all, so it evidently was not nearly as bright as a normal supernova. One theory is that the star that exploded had evaporated most of its outer layers before exploding. Chandra's first images (there will be other, much longer Chandra observations of Cas A) will allow scientists to test this theory. X-ray spectra taken with ACIS will make it possible to identify which heavy elements are present and in what quantities. These observations should help astronomers resolve the long-standing mystery as to the nature of Cas A.

A related mystery is whether the explosion that produced Cas A left behind a neutron star, black hole, or nothing at all. A bright point-like object near the center of the Chandra image may be the long-sought neutron star or black hole left behind in the explosion that produced Cas A. Longer observations of the central bright spot should reveal whether the spot is a neutron star, a black hole, a clump of hot gas, or an unrelated foreground or background object.

The study of remnants of exploded stars, or supernovas, is essential for our understanding of the origin of life on Earth. The cloud of gas and dust that collapsed to form the sun, Earth, and other planets was composed mostly of hydrogen and helium, with a small amount of heavier elements such as carbon, nitrogen, oxygen, and iron. The only place where these and other heavy elements necessary for life are made is deep in the interior of stars that are more than 10 times as massive as the sun. Supernovas are the creative flashes that renew the galaxy. They seed the interstellar gas with heavy elements, heat it with the energy of their radiation, stir it up with the force of their blast waves, and cause new stars to form. Chandra's unique sensitivity allows it to make images of many supernova remnants in our galaxy, and in neighboring galaxies as well.

The supernova remnant E0102–72 in the Small Magellanic Cloud, a nearby galaxy that is 190,000 light years from Earth, is believed to have resulted from the explosion of a massive star much like the one that produced Cas A. The Chandra image shows a source that resembles a flaming cosmic wheel stretching across 40 light years of space (Plate II). An analysis by Terry Gaetz of Harvard-Smithsonian and his colleagues showed that, like Cas A, E0102–72 has a bright inner ring surrounded by a less intense outer ring. The larger diameter of E0102–72 suggests that it is about three times as old as Cas A, a supposition that is supported by the lower temperatures—several million degrees compared to several tens of millions in Cas A.

Claude Canizares and his team at MIT used ACIS and the High-Energy Transmission Grating Spectrometer to study the source in detail.

They found that different parts of the bright inner ring are moving at different speeds, consistent with the double-shock-wave picture described above for Cas A. They also determined that the ring is exceptionally rich in oxygen, confirming the theory that massive stars provide most of the oxygen in the universe.

"In that regard, they [massive stars] can be called the fountains of life," Canizares said. "There is enough oxygen in that ring to mix with other materials and create about a thousand solar systems."

Another object that is the remnant of the explosion of a massive star is N132D, which is in the Large Magellanic Cloud. The Chandra image, made with the High-Resolution Camera (HRC), shows a highly structured shell of 10-million-degree gas that is 80 light years across. The remnant is thought to be about 3,000 years old. The Large Magellanic Cloud, a companion galaxy to the Milky Way, is 180,000 light years from Earth. In the Chandra image, the regions of brightest X-ray emission are shown in white. Despite its larger size and presumably older age, it has a higher temperature than E0102–72, suggesting that it was created in a more energetic explosion by a more massive star.

Detailed study of the Chandra images of N132D should allow astronomers to accurately determine the energy of the outburst and compare its development over time with that of Cas A, E0102–72, and other supernova remnants. The environment of N132D probably plays an important role in its evolution. Radio observations indicate that it is colliding with a giant cloud of dust and gas. The collision, which produces the brightening on the southern rim of the remnant, will slow the expansion of the shock wave. The relatively weak X-radiation on the upper left shows that the shock wave is expanding into a less dense region on the edge of the giant cloud.

The prelude to a supernova begins when a massive star has used up its nuclear fuel and the pressure drops in the central core of the star. The matter there is crushed to higher and higher densities until a neutron star, about 12 miles in diameter, is formed. The rest of the star continues to fall inward toward the neutron star. Temperatures rise to hundreds of millions, then billions of degrees Celsius. Out-of-control nuclear reactions pump a tremendous amount of energy into the infalling material. In a matter of seconds, the avalanche of infalling matter is transformed into a

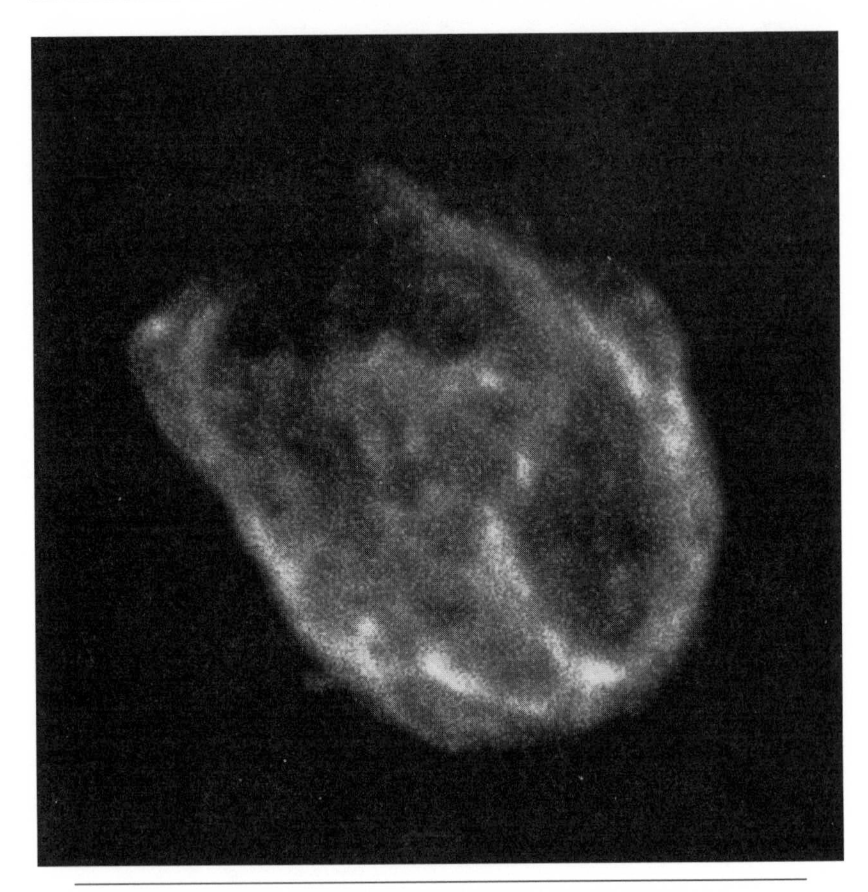

Chandra image of supernova remnant N132D. (NASA/SAO/CXC.)

runaway explosion. A thermonuclear shock wave races through the now expanding stellar debris, fusing lighter elements into heavier ones and producing a brilliant visual outburst that lasts for a few months.

As X-ray images from Chandra and other X-ray observatories have shown, the expanding stellar debris creates a shell of multimillion-degree gas that can be observed for thousands of years. One of the earliest discoveries of X-ray astronomy was the observation that the neutron star left behind after a supernova can also be a powerful source of X-radiation. The Crab Nebula, the remnant of a supernova explosion that was observed on Earth in 1054, was detected as a strong X-ray source by rocket-borne detectors in the early 1960s. In 1968 it was determined that the nebula contains a rapidly spinning neutron star that was observed to emit periodic flashes of radiation 30 times a second. The Crab neutron star is one of a class of objects called pulsars that were first observed with radio telescopes. The name pulsar is misleading because the flashes are not caused by pulsations of the neutron star; rather they are thought to be due to a rotating lighthouse-like beam from a rapidly rotating neutron star.

The flashing neutron star, or pulsar, is surrounded by a bright diffuse cloud, or nebula. The nebula, which is about 6 light years across, is expanding outward at 2 million miles per hour. The filamentary system visible in optical images is near the outer boundary of this expansion. Both the nebula and the pulsar are bright sources of radiation from radio through gamma ray energies. The primary source of this radiation is the central neutron star.

When the core of the star collapsed to form a neutron star, it spun up in much the same way that a twirling ice skater spins faster when she pulls in her arms, with one difference: the skater increases her rate of spin about tenfold, whereas the neutron star spins up to a hundred million times faster. The magnetic field increases by an even greater amount.

The rotating highly magnetized neutron star acts like a generator of awesome power. It creates electrical voltages 30 million times the voltage of a lightning bolt, and produces a total power equal to 100,000 times the power output by the sun. The neutron star consumes about 5 percent of this prodigious power output in the "lighthouse" radiation. The rest is transmitted to the nebula, making the Crab neutron star the most efficient power generator known.

The Chandra X-ray Observatory image of the Crab Nebula traces the flow of the energetic particles and gives new insight into the puzzle as to exactly how the neutron star transmits power to the nebula (Plate III). The central neutron star is surrounded by tilted rings, or waves, of high-energy particles that appear to have been flung outward over a distance of more than a light year from the pulsar. These rings could be the location of extremely high energy shock waves whose energy is being dumped into the nebula. Jet-like structures produced by high-energy particles blasting away from the neutron star in a direction perpendicular to the rings are also visible. The X-rays from the Crab Nebula are produced by high-energy particles spiraling around magnetic field lines in the nebula. The bell-shaped appearance of the nebula could be due to the interaction of this huge magnetized bubble with clouds of gas and dust in the vicinity.

The process of conversion of rotational energy into high-energy particles is one of the most efficient ways to generate energy from a gravitational field. The Crab Nebula provides astrophysicists with a nearby cosmic laboratory in which to study this important phenomenon that could have significance for many other problems in astrophysics. Extraordinary as it is, the Crab Nebula is, fortunately, not unique. Otherwise, any theory developed to explain what is observed there would be open to criticism that it had been developed to explain just one unusual object.

The Vela pulsar (Plate IV) shows a remarkably similar set of jets and rings; and the pulsar in PSR 0540–69, located in the Large Magellanic Cloud, has many features in common with the Crab neutron star. It emits pulses of radio, optical, and X-radiation at a rate of 50 per second, close to the Crab neutron star's rate of 30 per second. It has an age comparable to that of the Crab neutron star and a comparable total energy output. As with the Crab neutron star, only a few percent of the total energy output of the neutron star powerhouse is in the form of pulses. The rest is transmitted to a nebula around it. And also as with the Crab, the first Chandra image of PSR 0540–69 revealed a jet structure. This suggests that the process for transferring the power from the pulsar to the nebula is similar to the process in the Crab Nebula.

Cas A, the Crab Nebula, and the other supernova remnants discussed here are all thought to have originated in the explosion of a star with a mass between 10 and 30 times that of the sun. What happens when a star much more massive than this uses up its energy resources and collapses? Then, general opinion holds, it will explode and leave behind a black hole. Some astrophysicists have suggested that the most massive stars may produce titanic explosions called hypernovas that are a hundred times more energetic than a supernova, and are responsible for the mysterious gamma ray bursts that outshine everything in the universe at their peak.

One candidate for such an outburst is Eta Carinae, which is fortunately 7,500 light years away from Earth. Even so, if Eta Carinae were to become a hypernova or even a supernova, it would be one of the most spectacular cosmic events in recorded history. Until that time, Eta Carinae provides astronomers with an enigmatic and intriguing object. Between 1837 and

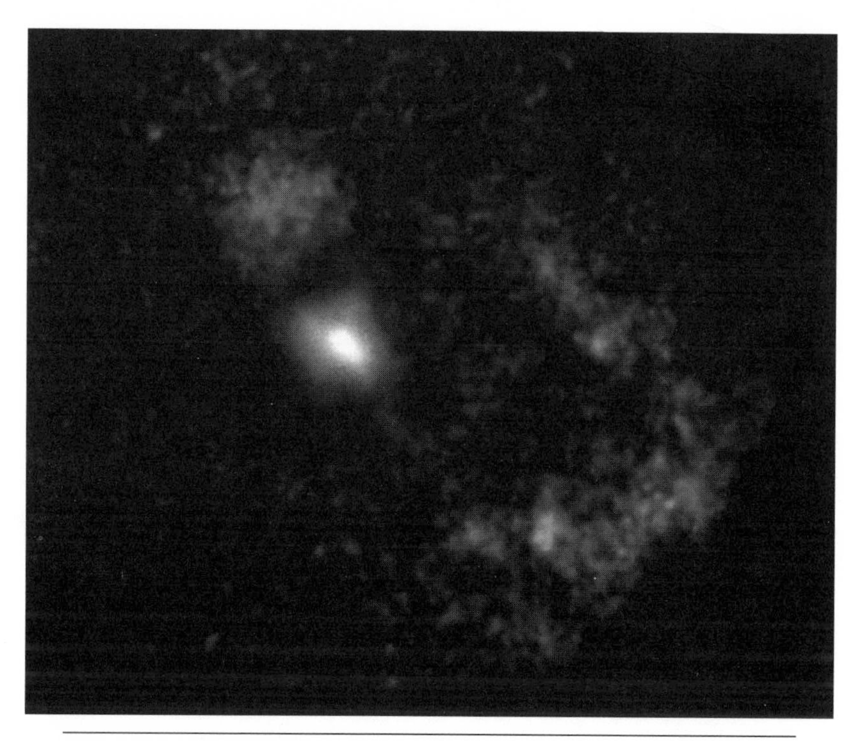

Chandra image of the supernova remnant PSR 0540–69. (NASA/SAO/CXC.)

1856 it increased dramatically in brightness to become the brightest star in the sky except Sirius, even though it is more than 800 times farther away from us than Sirius. This "Great Eruption," as it is called, had an energy comparable to a supernova, yet did not destroy the star, which faded to become a dim star, invisible to the naked eye. Since 1940, Eta Carinae has begun to brighten again, becoming visible to the naked eye.

Modern-day observations of Eta Carinae have shown it to be the most luminous object known in our galaxy. It radiates at a rate of several million times that of the sun. Most of the radiation is at infrared wavelengths, from dust in the bipolar nebula. Astronomers still do not know what lies at the heart of Eta Carinae. Most believe that it is powered by an extremely massive star that may have a companion. The massive star may be a hundred times as massive as the sun. Such stars produce intense amounts of radiation that cause violent instabilities before they explode as supernovas.

Chandra's X-ray image of Eta Carinae reveals a hot inner core around this mysterious superstar (Plate V). Three distinct structures can be seen: an outer, horseshoe-shaped ring about 2 light years in diameter, a hot inner core about 3 light months in diameter, and a hot central source less than a tenth of a light year in diameter that may contain the superstar.

All three structures are thought to represent shock waves produced by matter rushing away from the superstar at supersonic speeds. The temperature of the shock-heated gas ranges from 60 million degrees Celsius in the central regions to 3 million degrees Celsius on the outer edge.

An optical image of Eta Carinae made by the Hubble Space Telescope reveals two spectacular bubbles of gas expanding in opposite directions away from a central bright region at speeds in excess of a million miles per hour. The central region visible in the Chandra image has never been resolved before, and appears to be associated with a central disk of high-velocity gas rushing out at much higher speeds perpendicular to the hourglass-shaped optical nebula.

The Chandra X-ray image gives a glimpse deep into the nebula, where the fastest material being thrown off by Eta Carinae is found. The outer ring provides evidence of another large explosion that occurred more than a thousand years ago. Further Chandra observations of Eta Carinae are planned and should give astronomers deeper insight into this cryptic colossus.

If Eta Carinae does produce a black hole when it explodes, the black hole will have the mass equivalent of about 50 suns. This would make it larger than any known or suspected black hole in our galaxy, with one exception: the center of our galaxy, which is believed to harbor a black hole that has the mass of more than a million suns.

One of the most fascinating results to come out of the last decade of astrophysical research is the accumulation of compelling evidence that the central core of virtually every galaxy contains a supermassive black hole. If these giant black holes are given a sufficient supply of matter to swallow, they will produce enormous amounts of energy as gravitational energy is transformed into other forms. For example, on Earth hydroelectric plants convert the energy of falling water into electricity. In black holes, this process is carried to the extreme. The infall of the mass equivalent of one mosquito, about 1/100th of a gram, into a black hole every second would liberate the power of 100 Hoover dams!

Some of the most intense X-ray sources in the universe are caused by gas that is swirling toward a black hole. As the tremendous gravity of a black hole pulls gas and dust particles toward it, the particles speed up. Collisions between the infalling particles heat them to temperatures of many millions of degrees. Matter at such temperatures radiates primarily in X-rays, so Chandra is an efficient black hole probe.

Two Chandra observations of giant black holes in the centers of galaxies have yielded surprising results. In our Milky Way galaxy the central source is very weak in X-rays (Plate VIA). The giant black hole in the center of the Andromeda galaxy is surrounded by gas at an unexpectedly "cool" million degrees (Plate VIB). Current theories predict that the temperature of the gas falling into the central black hole in Andromeda should be greater than 10 million degrees Celsius.

"This observation is a challenge for our theoretical friends," said Stephen Murray, who along with Mike Garcia leads the Harvard-Smithsonian team that made the Andromeda observations.

Eliot Quataert an expert on black holes from the Institute for Advanced Study in Princeton, New Jersey, agrees. "The Chandra observation is telling us that an entirely different flow pattern is operating around the Andromeda black hole," he said. One possibility is that the matter near the black hole is undergoing a rapid boiling motion. This motion could allow the material to get rid of some of its excess energy before making its final descent into the black hole.

The giant black holes in the center of the Andromeda and Milky Way galaxies are like sleeping giants compared to those in some galaxies. The key difference is the amount of gas available for the black hole to swallow. If there is a generous supply, the power generated by the black hole can equal that of billions, even trillions of stars.

At the "nearby" distance of 11 million light years, the galaxy Centaurus A (Cen A) gives astronomers an excellent opportunity to study an active giant black hole, or active galactic nucleus, as astronomers call it. The energy output of the central black hole can in many cases affect the appearance of entire galaxies. In extreme cases, called quasars, the active nucleus can be a thousand times brighter than the host galaxy. The output from the Cen A black hole is slightly less than that of the entire host galaxy, so it is considered a relatively weak active nucleus. However, it is much closer to us than any quasar, so it is special.

The black hole in Cen A is believed to be of modest size, containing a mass equal to that of 10 million suns. One explanation for the explosive activity of Cen A's central black hole is that a collision with one or more smaller galaxies has provided a generous supply of gas for the supermassive black hole to accrete. The dust lanes seen in optical images that stretch across the middle of the galaxy may be remnants of such a collision.

One of the most intriguing features of supermassive black holes is that they do not suck up all the matter that falls within their sphere of influence. Some of the matter falls inexorably toward the black hole, and some explodes away from the black hole in high-energy jets that move at a speed near the speed of light. Radio observations of Cen A reveal a pair of oppositely directed jets bursting away from the central region.

The Chandra X-ray image of Cen A shows a strong X-ray source in the nucleus of the galaxy, at the location of the suspected supermassive black hole (Plate VIC). The bright jet extending out from the nucleus to the upper left is due to explosive or highly energetic activity around the black hole that ejects matter at high speeds from the vicinity of the black hole. A fainter "counter jet" extending to the lower right can also be seen. This jet is probably pointing away from us, which accounts for its faint appearance. The presence of a bright X-ray jet means that electric fields are continually accelerating electrons to extremely high energies over enormous distances. Exactly how this happens is a major puzzle that Chandra may help to solve.

X-ray jets have been observed in other galaxies, but none of them matches the jet discovered by Chandra in one of the first images it made. Chandra's X-ray image of the quasar PKS 0637–752, a quasar so distant that we see it as it was 6 billion years ago, showed a jet coincident with a radio jet that stretches over several hundred thousand light years (Plate VID). This is the largest known X-ray jet, and as such will place severe restrictions on any theory that seeks to explain how the energy of the jet is converted into high-energy particles so far from the central source.

The jet in the radio galaxy 3C295 is not nearly so large as the one in PKS 0637–752, even though the central source is just as powerful. The reason for this is most likely its environment. It sits at the center of a vast cloud of multimillion-degree gas that is several million light years across. The pressure of this surrounding gas cloud has probably slowed the jet down, in much the same way that jets of water in a jacuzzi are slowed down by the pressure of the water in the tub. The gas cloud around 3C295 contains hundreds of other galaxies swarming like bees around a hive. This concentration of galaxies and gas, of which 3C295 is a part, is called a galaxy cluster. At a distance of 7 billion light years, 3C295 is one of most distant clusters that has been observed in X-rays. The Chandra image showed for the first time that the jet and the nucleus of the central galaxy are both emitting large quantities of X-rays from extremely high-energy electrons (Plate VIIA). The pressure in the hot gas is enough to slow but not stop the jet. The jet will probably expand to many times its present size of 30,000 light years, which is already comparable to the size of our Milky Way galaxy.

Chandra's image of the Hydra A galaxy cluster has revealed a possible way of explaining the fate of the largest objects in the universe (Plate VIII). For years astronomers have been searching unsuccessfully for large quantities of matter they believed must be flowing into the central regions of galaxy clusters. The Chandra image of Hydra A displays for the first time long snake-like strands of 35-million-degree gas extending away from the center of the cluster. These structures show that the inflow of cooling gas is deflected by magnetic fields produced by explosions from a central black hole.

The X-ray image also reveals a bright wedge (shown in white) of hot, multimillion-degree gas pushing into the heart of the cluster. Like the legendary Hercules, who had to contend with the multiple heads of the monstrous Hydra, astrophysicists now know they must deal with the effects of magnetic fields, star formation, rotation, and black holes if they are to understand what is happening in the inner regions of the galaxy cluster.

Previous X-ray observations indicated that the gas in the inner regions of Hydra A should be cooling and slowly settling into the center of the cluster to form new galaxies or hundreds of trillions of dim stars. As astronomers began searching for this cool matter, they were puzzled to find that the new galaxies and stars were not detected in sufficient numbers.

Chandra's spectacular image of Hydra A, which is 840 million light years from Earth, may point to a resolution of this problem. The inflow of cooling gas may be deflected by magnetic fields, and even pushed back into the cluster by explosions from the vicinity of a supermassive black hole at the core of the central galaxy.

In Hydra the whole cycle can be observed. There is the hot gas cloud, the disk of material feeding the black hole, and evidence from both radio and X-ray observations that the explosion from the gas near the black hole is pushing the hot gas around. A vast bubble of high-energy particles is apparently pushing the hot gas aside, creating the Hydra-like loops of hot gas. Similar processes are likely to be at work in other galaxy clusters, and in newly forming galaxies that are collapsing from a cloud of gas.

Galaxy clusters grow to vast sizes as smaller clusters are pulled inward under the influence of gravity. They collide and merge over the course of billions of years, releasing tremendous amounts of energy that heats the cluster gas to 100 million degrees Celsius. The Chandra image of the galaxy cluster Abell 2142 provides the first detailed look at the late stages of this merger process (Plate VIIB). Previous X-ray images from the Roentgensatellite (Rosat) suggested that two clouds were in the process of coalescing into one, but the details remained unclear. Chandra measured variations of temperature, density, and pressure with unprecedented resolution.

The result is a picture of a colossal cosmic "weather system" produced by the collision of two clusters of galaxies. For the first time, the pressure fronts in the system can be traced in detail, and they show a bright, but relatively cool 50-million-degree central region embedded in a large elongated cloud of 70-million-degree gas, all of which is roiling in a faint "atmosphere" of 100-million-degree gas.

"We can compare this to an intergalactic cold front," said Maxim Markevitch of Harvard-Smithsonian and leader of the international team involved in the analysis of the observations. "A major difference is that in this case, cold means 70 million degrees."

The merging gas clouds are in the core of Abell 2142. This cluster is 6 million light years across and contains hundreds of galaxies and enough gas to make a thousand more. It is one of the most massive objects in the universe.

"Now we can begin to understand the physics of these mergers, which are among the most energetic events in the universe," said Markevitch. "The pressure and density maps of the cluster show a sharp boundary that can only exist in the moving environment of a merger."

With this information scientists can make a comparison with computer simulations of cosmic mergers. This comparison, which is still in the early stages, shows that this merger has progressed to an advanced stage. Strong shock waves predicted by the theory for the initial collision of clusters are not observed. It appears likely that these subclusters have collided two or three times in a billion years or more, and have nearly completed their merger.

Galaxy clusters are formed through the merger of smaller groups and clusters of galaxies over billions of years. Because of the distances involved, millions of light years, a collision between galaxy clusters occurs on a time scale of hundreds of millions of years. The age of the dinosaurs could have come and gone on Earth in the time it takes for one such collision to happen.

The X-rays from the most distant X-ray sources have been traveling through space since before the Earth was formed 4 and 1/2 billion years ago. Some of the X-rays may be more than 10 billion years old. These X-rays come from sources so distant that, before Chandra, they were too faint to detect individually. They showed up as part of a diffuse background glow of X-rays.

The X-ray background glow was detected on a rocket flight in 1962 when Riccardo Giacconi and his colleagues discovered the first X-ray source outside our solar system. Some scientists suspected that the X-ray background was due to extremely distant active galaxies or quasars. They were unable to prove this, however, for no X-ray telescope until Chandra has had both the sharp vision and the sensitivity to detect the individual sources.

In January 2000 two teams of astronomers, one led by Richard Mushotzky of NASA's Goddard Space Flight Center, and another led by Gordon Garmire of Penn State University, announced that they may have cleared up the background puzzle.

"Chandra was designed in some sense to solve this problem," Mushotzky said. "It has the sensitivity, and it has the angular resolution, so it can go very faint [detect very faint sources] to find the objects that produce the X-ray background. Because it provides accurate positions, you can follow these up in other wavelength bands and figure out what the sources are."

Using Chandra, Mushotzky's team resolved over 70 percent of the background glow into individual sources, about two thirds of which are quasars or galaxies with a bright central core that is presumably due to matter falling into a giant black hole. The nature of the remaining sources is still a mystery, though.

"The mystery of the X-ray background is well on its way to being resolved," said Garmire, whose group obtained similar results. "But there are new mysteries coming to the fore in the form of objects that we can't identify with any known object."

What are these objects? They could be galaxies so distant that the optical light gets absorbed by cool gas during its long journey across the universe. Or galaxies that are completely surrounded by dust. Or galaxies that have yet to form stars and are still just clouds of hot gas with a central giant black hole. Or some entirely new phenomena.

As the circle of knowledge expands into the unknown, the boundary between the known and the unknown also expands. Marvelous endeavors like Chandra are pushing the boundary outward. Speculating and predicting what lies beyond the boundary is fascinating. Finding out is even more fascinating.

Bibliographical Notes

INTRODUCTION

Quotes from Galileo's *The Starry Messenger* are as quoted in D. Boorstein, *The Discoverers* (New York: Random House, 1983), p. 320.

1. HIGH-ENERGY VISION

E. Segre's quote is from E. Segre, *From X-rays to Quarks* (New York: Freeman, 1980), p. 242.

B. Rossi's quote is from B. Rossi, *X-ray Astronomy*, ed. R. Giacconi and H. Gursky (Dordrecht, Holland: D. Reidel, 1971), p. vii.

R. Giacconi's quote is from W. Tucker and K. Tucker *The Cosmic Inquirers* (Cambridge: Harvard University Press, 1986), p. 52.

2. INVISIBLE LIGHT

Herschel's quote is available in H. King, *The History of the Telescope* (New York: Dover, 1955), p. 140; quoted here from W. Herschel in J. Dreyer, ed., *Scientific Papers of Sir William Herschel* (1912), p. 63.

M. Faraday's quote is from M. Faraday, *Experimental Research in Electricity* (New York: Dover, 1965), vol. 3, p. 19.

3. LIGHT QUANTA

M. Faraday's quote is from M. Faraday, *Experimental Research in Electricity* (New York: Dover, 1965), vol. 3, p. 19.

Quotes from Maxwell on Faraday are from Maxwell, *A Treatise on Electricity and Magnetism* (New York: Dover, 1954), vol. 2, p. 176.

R. Feynmann's quote is from R. Feynmann, *Lectures on Physics* (Reading, MA: Addison-Wesley, 1964), vol. 2, p. 69.

Einstein and Infield's quote is from A. Einstein and L. Infeld, *The Evolu-*

tion of Physics from Early Concepts to Relativity and Quanta (New York: Simon & Schuster, 1938).

E. Segre's quote is from E. Segre, *From X-rays to Quarks* (New York: Freeman, 1980), p. 86.

4. THE BIRTH OF X-RAY ASTRONOMY

Basic references for the early history of X-ray astronomy are R. Giacconi and H. Gursky, eds., *X-ray Astronomy* (Dordrecht, Holland: D. Reidel, 1974), and W. Tucker and R. Giacconi, *The X-ray Universe* (Cambridge: Harvard University Press, 1985).

5. X-RAY STARS

For a readable and authoritative treatment of white dwarfs, neutron stars, and black holes, see K. Thorne *Black Holes and Time Warps* (New York: W. W. Norton, 1994), p. 219ff. For a discussion of the early work on the formation of black holes, see D. Arnett, *Supernovae and Nucleosynthesis* (Princeton: Princeton University Press, 1995). For a discussion of the early observations and theoretical modeling, see W. Tucker and R. Giacconi, *The X-ray Universe* (Cambridge: Harvard University Press, 1985).

J. Oppenheimer and H. Snyder's quote is from *Physical Review*, 56 (1939), 455.

6. THE UHURU YEARS

For discussions of the Uhuru mission see W. Tucker and R. Giacconi *The X-ray Universe* (Cambridge: Harvard University Press, 1985); W. Tucker and K. Tucker, *The Cosmic Inquirers* (Cambridge: Harvard University Press, 1986); and R. Giacconi and H. Gursky, eds., *X-ray Astronomy* (Dordrecht, Holland: D. Reidel, 1974).

Quotes from M. Oda and his colleagues are from M. Oda et al., *Astrophysical Journal Letters*, 166 (1971), L1.

R. Giacconi's quotes are from Tucker and Tucker, *The Cosmic Inquirers*.

7. THE EINSTEIN OBSERVATORY

More details on the building of and the scientific results from the Einstein Observatory can be found in W. Tucker and R. Giacconi, *The X-ray*

Universe (Cambridge: Harvard University Press, 1985), and W. Tucker and K. Tucker, *The Cosmic Inquirers* (Cambridge: Harvard University Press, 1986).

Giacconi's move from AS&E is described in Tucker and Giacconi, *The X-ray Universe,* p. 110.

The quotations are from Tucker and Tucker, *The Cosmic Inquirers,* pp. 81–89.

8. READY FOR THE JOB

Quotations are from an interview with C. Pellerin in Boulder, CO, on 9/8/95.

The fallout from Hubble overruns is discussed in R. W. Smith, *The Space Telescope* (Cambridge: Cambridge University Press, 1989), p. 312.

9. BLAZING THE TRAIL

Quotes from H. Tananbaum are from an interview on 8/10/94 in Cambridge, MA.

Quotes from the report of the Allen Board of Inquiry are from E. Chaisson, *The Hubble Wars* (New York: HarperCollins, 1994), p. 224.

Quotes from F. Wojtalik are from an interview on 6/5/97 in Huntsville, AL.

Quotes from M. Weisskopf are from an e-mail dated 4/24/98.

The report of the AXAF science working group appeared in NASA Technical Memorandum 788285, "Advanced X-ray Astrophysics Facility (AXAF)—Science Working Group Report," May 1980, p. vi.

The Astronomy Survey Committee quote is from *Astronomy and Astrophysics for the* 1980's, vol. 1: *Report of the Survey Committee* (Washington, DC: National Academy Press, 1982), p. 27.

10. JOCKEYING FOR POSITION

Quotes from R. Giacconi are from A. Finkbeiner, "Studies from Life," *The Sciences,* May/June 1993.

S. Murray's quote is from an interview in June 1982 in Cambridge, MA.

M. Weisskopf's quotes are from a 4/16/98 e-mail.

M.Zombeck's quote is from a 4/24/98 e-mail.

H. Tananbaum's quotes are from an 8/10/94 interview.

C. Pellerin's quotes are from a 9/8/95 interview.

11. THE GREAT OBSERVATORIES

C. Pellerin's quotes are from a 9/8/95 interview.

M. Harwit's quote is from M. Harwit, *Cosmic Discovery* (New York: Basic Books, 1981), p. 43.

12. THE BELMONT RETREAT

C. Pellerin's quotes are from a 9/8/95 interview.

G. Keyworth's quote is from a letter from Keyworth to J. Beggs, 6/4/84.

13. PROGRESS AND SETBACKS

C. Pellerin's quotes are from a 9/8/95 interview.

C. Canizares's quotes are from an interview on 8/1/96 in Cambridge, MA.

The account of Giacconi's meeting with Reagan is taken from a letter of Giacconi to NASA Administrator J. Fletcher dated 8/14/96.

D. Schwartz's quotes are from an interview on 11/12/98 in Cambridge, MA.

L. van Spebroeck quotes are from an interview on 8/2/94 in Cambridge, MA.

NASA Administrator J. Beggs's legal troubles were reported by E. Marshall, *Science*, 237 (1987), 21, and in *Time*, Nov. 20, 1989.

14. THE CHALLENGE IS SET

C. Pellerin quotes are from an interview on 9/8/95.

L. Fisk quotes are from an interview on 12/12/95 in San Francisco.

H. Tananbaum quotes are from an interview on 7/29/94 in Cambridge, MA.

R. Malow quotes are from an interview on 8/8/96 in Cambridge, MA.

Quotes about Malow and the House Appropriations Committee are from R. Munson, *The Cardinals of Capital Hill* (New York: Grove Press, 1993).

15. THE CHALLENGE IS MET

Quotes from H. Tananbaum are from an interview on 7/29/94 in Cambridge, MA.

Quotes from A. Napolitano, I. Schmidt, P. Reid, M. Magida, and T. Gordon are from interviews in Danbury, CT, on 8/17/94.

The description of the precision metrology mount is from L. Cernoch et al., *Proc. SPIE*, 133 (1990), 374.

Quotes from R. Malow are from an interview on 8/8/96 in Cambridge, MA.

Quotes from G. Matthews, T. Casey, and M. Freeman are from interviews in Rochester, NY, on 8/30/96.

Quotes from J. Hughes are from an interview in Cambridge, MA, on 8/10/94.

Quotes from L. Van Speybroeck are from an interview in Cambridge, MA, on 8/2/94.

Quotes by H. Tananbaum and R. Malow as reported by E. Marshall are from E. Marshall, *Science,* 254 (1991), 510.

Quotes from L. Fisk are from a 12/12/95 interview.

Quotes from F. Wojtalik are from an interview on 6/5/97.

16. A BRUISING LESSON

Quotes from R. Traxler and his aides and from C. Schumer and R. Truly are from R. Munson, *The Cardinals of Capitol Hill* (New York: Grove Press, 1993).

Quotes from J. Grindlay and K. Bailey are from their Space Science Working Group memorandum dated 9/30/91.

Quote from K. Lestition is from an interview in Cambridge, MA, on 4/23/99.

L. Fisk's quotes are from a 12/12/95 interview.

C. Pellerin's quotes are from a 9/8/95 interview.

H. Tananbaum's quotes are from a 7/20/94 interview in Cambridge, MA.

M. Weisskopf's quotes are from H. Tananbaum's notes of a 2/25/92 Chandra science working group meeting.

17. CONFLICT AND COMPROMISE

R. Truly's removal as NASA Administrator was reported in *U.S. News & World Report,* Feb. 24, 1992, and in *Science,* 255 (1992), 915.

H. Tananbaum's quotes are from a 7/20/94 interview in Cambridge, MA.

C. Pellerin's quotes are from a 9/8/95 interview.

L. van Speybroeck's quotes are from an 8/2/94 interview in Cambridge, MA.

R. Giacconi's quote is from a letter to the AXAF science working group, 4/6/92.

R. Schilling's quotes are from a 10/30/97 interview in Redondo Beach, CA.

C. Canizares's quote is from an 8/1/96 interview.
R. Malow's quotes are from an 8/8/96 interview.
L. Fisk's quotes are from a 12/12/95 interview.
The departure of L. Fisk from NASA was reported in *American Astronomical Society Newsletter*, 65 (June 1993).

18. GRINDING, POLISHING, AND COATING THE MIRRORS

For general background on the mirrors, see P. Reid, "AXAF Optics Fabrication–Convergence and *cmi*," *TechNews*, 6 (Feb. 1995), 1–5, and A. Burr and M. Magida, "Completion of AXAF Mirrors—A Major Milestone in HDOS History," ibid., p. 5.
Quotes from R. Hahn, J. Johnston, and R. Langley are from 12/11/95 interviews in Santa Rosa, CA.

19. THE MIRROR ASSEMBLY

L. Cohen's remarks are from an interview in Cambridge, MA, on 7/2/99.
All other quotes in this chapter are from interviews in Rochester, NY, on 8/30/96.

20. CALIBRATION

Quotes in this chapter are from interviews with D. Schwartz in Cambridge, MA, on 8/19/97, C. Jones in Cambridge, MA, on 8/29/97, H. Tananbaum in Cambridge on 8/26/97, and M. Weisskopf in San Diego on 6/5/98.

21. THE SCIENTIFIC INSTRUMENTS

Quotes in this chapter are from interviews with S. Murray in Cambridge, MA, on 8/23/96 and 8/20/97, with C. Canizares in Cambridge on 8/1/96, with G. Garmire in San Diego on 5/1/96, with C. Canizares and Kathy Flanagan in Cambridge on 5/2/99, and with H. Tananbaum in Cambridge on 8/26/96 and 8/26/97. P. Predhel's quotes are from an e-mail on 4/18/99, and M. Bautz's quotes are from a 5/5/00 e-mail.

22. AN OBSERVATORY AND A NAME

Quotes from J. Payne, B. McKinney, J. Korka, S. Loer, R. Schilling, and E. Wheeler are from interviews in Redondo Beach, CA, on 10/30/97.

Quotes from F. Wojtalik are from NASA press releases on 12/5/97 and 10/13/98, and from an interview at the TRW media event at Redondo Beach on 4/20/98.

C. Staresinich's quotes are from interviews in Cambridge, MA, on 8/10/99 and 8/11/99.

E. Collins's quote is from her comments at the TRW media event in Redondo Beach on 4/20/98.

G. Austin's quote is from a conversation at the Wok and Roll restaurant in Cambridge, MA, on 7/7/98.

H. Tananbaum's quote "They ruled out" is from an interview in Cambridge, MA, on 2/11/99.

G. Garmire's quote is from *Chandra News*, 6 (Feb. 1999), 5.

H. Tananbaum's quote "They are making good progress" is from a private conversation.

E. Weiler's quote "Delaying the shipment" is from *Space News*, Jan. 26, 1999.

H. Tananbaum's quote "I'm through" is from a conversation at the Wok and Roll restaurant on 10/28/98.

K. Ledbetter's quote is from a telephone press conference on 1/20/99.

H. Tananbaum's quotes on complexity are from a 2/11/99 interview.

H. Tananbaum's and L. van Speybroeck's quotes are from a conversation at the Wok and Roll restaurant on 4/23/99.

D. Goldin's quote is from Florida Today Space Online, June 7, 1999.

23. LAUNCH

L. Chandrasekhar's quotes are from the prelaunch ceremony at the Kennedy Space Center on 7/19/99.

L. van Speybroeck's quote is from comments made on 8/25/99 in Cambridge, MA.

H. Tananbaum's quote is from a telephone conversation on 7/22/99.

24. ACTIVATION

C. Coleman's and E. Collins's quotes are from an in-orbit press conference on 7/26/99.

R. Brissenden's quote is from a conversation on 11/11/98 in Cambridge, MA.

Quotes from M. Rosenthal and others at the Chandra X-ray Center are from conversations W. and K. Tucker heard or participated in.

25. FIRST LIGHT

Quotes from people in the Action Room are from conversations W. and K. Tucker heard or participated in.

C. Canizares's quote on first light for the high-energy transmission grating is from an MIT press release dated 8/30/99.

S. Murray's and G. Austin's comments on the High-Resolution Camera's first light are from an interview on 9/3/99 in Cambridge, MA.

The calibration team's quotes are from a meeting on 9/13/99 in Cambridge, MA.

B. McNamara's discussions on the CTI problem are from e-mails in the period 9/9/99 to 10/11/99.

M. Bautz's quotes are from a 5/5/00 e-mail.

H. Tananbaum's quote is from a conversation on 10/8/99 in Cambridge, MA.

C. Pellerin's quote is from an interview on 9/8/95 in Boulder, CO.

L. van Speybroeck's quote is from a conversation on 9/27/99 in Cambridge, MA.

26. THE EXPLORATION BEGINS

E. Feigelson's quote is from an interview at the American Astronomical Society (AAS) meeting in Atlanta on 1/13/00.

C. Canizares's quote is from a press conference at the AAS meeting in Atlanta, 1/14/00.

S. Murray's quote is from a press conference at the AAS meeting in Atlanta, 1/14/00.

E. Quatert's quote is from an e-mail on 1/5/00.

M. Markevitch's quotes are from an e-mail on 2/21/00.

R. Mushotzky's quotes are from a press conference at the AAS meeting in Atlanta, 1/13/00.

G. Garmire's quotes are from a press conference at the AAS meeting in Atlanta, 1/14/00.

Acknowledgments

THIS BOOK is about people who wanted to know what lay beyond the horizon. Like successful adventurers of old, to reach their goal they had to lay elaborate plans, involve many people, spend buckets of gold, be politically savvy, persevere through long years of preparation, survive serious setbacks, be willing to risk it all in one glorious moment, and ultimately, fortune had to smile on them.

As with any story of this complexity, involving thousands of people and stretching over decades, we have surely missed many of the people and events that were crucial to its success. For this we apologize. We also regret that we were unable to present all the fascinating stories that we heard over the course of 6 years about the adventure of making NASA's Chandra X-ray Observatory a reality.

Chandra changed the lives of many people. Ours were no exception. A year before launch we joined the Chandra team to organize the Chandra public information office, Karen as a science writer, Wallace as science spokesman. Although we think that the final chapters of the story benefited from our involvement in the program, we also concede that our objectivity may have been affected. We have tried to minimize this by being mindful of any bias and by letting the participants tell the story in their own words as much as possible.

In the course of our research for this book, we interviewed a large number of people who had worked on Chandra over the years. We are grateful for their generosity in taking the time to talk with us about their experiences, and profoundly impressed by their dedication to the program. All the people we encountered, from technicians to program managers, were proud and honored to be part of a program of national importance. As citizens, we can be proud and honored that they were there to support, build, and operate a premier space observatory.

We thank Charles Atkinson, Gerry Austin, Mark Bautz, Dana Berry, Hale Bradt, Roger Brissenden, Rob Cameron, Claude Canizares, Tom Casey, Lester Cohen, Martin Elvis, Giuseppina Fabbiano, Yvette Femiano, Lennard Fisk, Kathleen Flanagan, Bill Forman, Mark Freeman, Gordon Garmire, Tom Gordon, Paul Gorenstein, Bob Hahn, Keith Havey, Jack Hughes, Jerry Johnston, Christine Jones, Ed Kellogg, Jim Korka, Kimberly Kowal, Bob Langley, Kathleen Lestition, Steve Loer, Mark Loucka, Matt Magida, Richard Malow, Gary Matthews, Brooks McKinney, Brian McNamara, Maxim Markevitch, Steve Murray, Art Napolitano, Joe Payne, Charles Pellerin, Larry Peterson, Charles Placito, Peter Predehl, Kathy Rapp, Paul Reid, Ralph Schilling, Dan Schwartz, Ira Schmidt, Fred Seward, David Sime, John Spina, Craig Stairsinich, Ed Swigonski, Martin Weisskopf, Fred Wojtalik, Jeff Wynn, Martin Zombeck, and many others for their assistance. We also thank Nancy Clemente and Michael Fisher for their encouragement and valuable editorial advice, and Sara Davis for her help with the illustrations and the permissions.

Special thanks are due to Leon van Speybroeck, who patiently answered countless questions, and to Harvey Tananbaum, who took the time on many occasions to give us the benefit of his unique and comprehensive knowledge of the Chandra program.

Index